MOLECULAR DYNAMICS

TOPICS IN MOLECULAR AND STRUCTURAL BIOLOGY

Series Editors

Stephen Neidle
Institute of Cancer Research
Sutton, Surrey, UK

Watson Fuller
Department of Physics
University of Keele, UK

Recent titles

Protein–Nucleic Acid Interaction
Edited by Wolfram Saenger and Udo Heinemann (1989)

Calcified Tissue
Edited by David W. L. Hukins (1989)

Oligodeoxynucleotides: Antisense Inhibitors of Gene Expression
Edited by Jack S. Cohen (1989)

Molecular Mechanisms in Muscular Contraction
Edited by John M. Squire (1990)

Connective Tissue Matrix, Part 2
Edited by David W. L. Hukins (1990)

New Techniques of Optical Microscopy and Microspectroscopy
Edited by Richard J. Cherry (1990)

Molecular Dynamics: Applications in Molecular Biology
Edited by Julia M. Goodfellow (1990)

MOLECULAR DYNAMICS

APPLICATIONS IN MOLECULAR BIOLOGY

Edited by

JULIA M. GOODFELLOW

Dept of Crystallography
Birkbeck College
London, UK

MACMILLAN
PRESS
Scientific & Medical

First published 1991

Published by
THE MACMILLAN PRESS LTD
Houndmills, Basingstoke, Hampshire RG21 2XS
and London
Companies and representatives
throughout the world

Filmset by Wearside Tradespools
Fulwell, Sunderland
Printed in Great Britain by
WBC Ltd., Bridgend

British Library Cataloguing in Publication Data
Molecular dynamics.
1. Molecules. Dynamics & structure
I. Goodfellow, Julia M. II. Series
539.6
ISBN 0-333-49886-0
ISSN 0265-4377

Contents

Preface

In the introduction to this series in 'Topics in Molecular Biology' in 1980, great emphasis was placed on the importance of molecular structure for understanding biological phenomena. At that time, the emphasis was on the three-dimensional models determined from experimental methods such as X-ray crystallography; 'more speculative model building and conformational analysis' was mentioned only in passing. In the decade since this was written, the original emphasis on molecular structure has increased with many new structures being solved.

Along with the accumulation of new three-dimensional structures, there has been an awareness that it is not the static picture alone but also the dynamics of macromolecules which is important. Although X-ray crystallography can model some aspects of atomic motion, it has been the extension of computer simulation techniques such as molecular dynamics from physics and chemistry to molecular biology which has provided an overall picture of macromolecular dynamics.

The aim of this book is to review the uses of molecular dynamics and related modelling techniques as applied to biological macromolecules. It is particularly suited for those who have not specialized in simulation techniques. As well as describing applications to proteins, nucleotides and carbohydrates, we have also reviewed the exciting possibilities which emerge from the use of free energy perturbation methods in protein engineering and drug design.

Recently, molecular dynamics has started to play a direct role in the determination of macromolecular structures themselves when simulated annealing protocols are used in combination with experimental data from X-ray crystallography or NMR spectroscopy.

I would like to thank all the authors for their contributions, my colleagues for their advice and David Grist of Macmillan Press for guiding me through the role of volume editor.

London, 1990 J.M.G.

The Contributors

Axel T. Brünger
The Howard Hughes Medical Institute and Department of Molecular Biophysics and Biochemistry
Yale University
New Haven
CT 06511, USA

Leo S. D. Caves
Department of Chemistry
University of York
Heslington
York YO1 5DD, UK

Tom P. Flores
Department of Crystallography
Birkbeck College
Malet Street
London WC1E 7HX, UK

Julia M. Goodfellow
Department of Crystallography
Birkbeck College
Malet Street
London WC1E 7HX, UK

Stephen C. Harvey
Department of Biochemistry
Schools of Medicine/Dentistry
University of Alabama
Birmingham
AL 35294, USA

Roderick E. Hubbard
Department of Chemistry
University of York
Heslington
York YO1 5DD, UK

Jutta Köhler
Genzentrum
MPI Biochemie
Am Klopferspitz
D-8033 Martinsried
FR Germany

Arun Malhotra
Department of Biochemistry
Schools of Medicine/Dentistry
University of Alabama
Birmingham
AL 35294, USA

David S. Moss
Department of Crystallography
Birkbeck College
Malet Street
London WC1E 7HX, UK

Dzung T. Nguyen
Department of Chemistry
Harvard University
12 Oxford Street
Cambridge
MA 02138, USA

Mansoor A. S. Saqi
Department of Crystallography
Birkbeck College
Malet Street
London WC1E 7HX, UK

Mark A. Williams
Department of Crystallography
Birkbeck College
Malet Street
London WC1E 7HX, UK

Robert K.-Z. Tan
Department of Biochemistry
Schools of Medicine/Dentistry
University of Alabama
Birmingham
AL 35294, USA

1
Simulating the Dynamics of Macromolecules

T. P. Flores and D. S. Moss

INTRODUCTION

Equilibria in biological systems are governed by their tendency to move towards states of lower free energy. The equilibrium constant K for a chemical reaction at constant temperature and pressure is related to the change in free energy by the equation

$$\Delta G = -RT \ln K \tag{1.1}$$

where G is the Gibbs free energy, R is the gas constant and T is the absolute temperature. The change in G may be related to the changes in the internal energy E, pressure P, volume V and entropy S by the equation

$$\Delta G = \Delta E + P\,\Delta V - T\,\Delta S \tag{1.2}$$

In a well-ordered system at modest temperatures ΔE may dominate this equation and the stable conformation of a molecular system may be calculated by minimizing the internal energy E. This energy is the sum of potential and kinetic energies and, as the latter depends only on temperature, the problem reduces to the minimization of the potential energy. This is discussed in the section entitled Other methods employed in molecular simulations.

In biological systems there is usually considerable disorder. The side groups on the surface of a protein usually exhibit much flexibility and the aqueous environment displays diffusive behaviour. The entropy term in Equation (1.2) is therefore significant and the determination of equilibria constants and equilibrium configurations must directly address the computationally more difficult task of determining free energy. This requires the simulation of a molecule in its aqueous environment using molecular dynamics (MD) or Monte Carlo (MC) methods as described in the sections on molecular dynamics and Monte Carlo simulation.

Motions in macromolecules cover a wide range of time scales. Vibrations of covalent bonds take place in 10^{-14} s, whereas large-scale conformational changes may take more than 10^2 s. It is this range of characteristic times that presents molecular simulation with one of its greatest challenges. The types of motions which are exhibited in protein molecules have been the subject of several reviews. Karplus and McCammon (1986) have given an overview of the different simulation techniques and their application to the different types of dynamic processes in protein molecules.

Most molecular dynamics studies of protein molecules have been based on classical equations of motion. The position and velocity of each atom have been assumed to be simultaneously defined. This assumption is reasonable for carbon atoms but a quantum mechanical treatment becomes necessary for lighter particles such as electrons and possibly protons. A path integral method of treating the quantum mechanical problem has been developed (Kuki and Wolynes, 1987; Zheng *et al.*, 1988) but this chapter will concentrate on the classical methods which can be applied to protein dynamics as long as proton and electron transfer are not relevant.

MOLECULAR DYNAMICS

The essential principle of MD is the numerical integration of the classical equations of motion for a system of interacting particles over a certain period of time. The trajectorics and velocities of the atoms are followed over this time period which is often limited by the computing resources available. One hour of processor time on a Cray X-MP is required for the simulation of 1 ps (10^{-12} s) of a small protein in its aqueous environment. The folding of a protein molecule into its native conformation may take several seconds and such studies are therefore not practicable. Most studies are restricted to time periods of less than 1 ns (10^{-9} s). The atoms are usually assigned initial velocities corresponding to a Maxwellian distribution at a given temperature. The simulation proceeds in a series of small time increments and, after each time step, the force on each atom is evaluated.

The force $F_i(t)$ on an atom i is obtained by taking the derivative of the potential energy function with respect to its position $r_i(t)$:

$$F_i(t) = -\frac{\partial}{\partial r_i} \mathcal{V}(r_1, r_2, \ldots, r_{3N_{\mathrm{at}}}) \tag{1.3}$$

The acceleration $a_i(t)$ of this atom with mass m_i is found by

$$a_i(t) = \frac{F_i(t)}{m_i} \tag{1.4}$$

and hence the atomic position

$$\frac{\mathrm{d}^2\boldsymbol{r}_i}{\mathrm{d}t^2}(t) = \frac{F_i(t)}{m_i} \tag{1.5}$$

It follows that for each atom there are three scalar equations which can be written in component form as

$$m_i a_i(t) = \sum_j f_{ij}(t) \tag{1.6}$$

The summation on the right-hand side is over all atoms j exerting forces on atom i which have to be evaluated. The computer solution of the differential equations (1.6) is carried out by replacing them with difference equations which are successively solved for small time steps, Δt. Consider the x coordinate of atom i in the following two difference equations, which are Taylor series approximations:

$$x_i(t+\Delta t) = x_i(t) + v_i(t)\,\Delta t + a_i(t)\frac{\Delta t^2}{2} \tag{1.7}$$

$$x_i(t-\Delta t) = x_i(t) - v_i(t)\,\Delta t + a_i(t)\frac{\Delta t^2}{2} \tag{1.8}$$

Adding Equations (1.7) and (1.8) gives

$$x_i(t+\Delta t) = 2x_i(t) - x_i(t-\Delta t) + a_i(t)\Delta t^2 \tag{1.9}$$

Substituting from Equation (1.6) then gives

$$x_i(t+\Delta t) = 2x_i(t) - x_i(t-\Delta t) + \sum_j \frac{f_{ij}(t)}{m_i} \tag{1.10}$$

Equation (1.10) gives us a way of predicting the position of atom i at time $t+\Delta t$ if we are given its positions at time t and $t-\Delta t$. Similarly, by subtracting Equations (1.7) and (1.8) we obtain

$$v_i(t) = \frac{x_i(t+\Delta t) - x_i(t-\Delta t)}{2\,\Delta t} \tag{1.11}$$

which gives the velocity at time t.

The equation (1.10) is known as the Verlet algorithm and is used to track the position of each atom at succeeding time steps (Verlet, 1967). Numerous other integration algorithms have been used, such as Gear, Leapfrog and Beeman. Some techniques, such as the Runge–Kutta methods, require several force evaluations per time step. These are unsuitable for molecular dynamics of proteins, where force evaluation is the most time-consuming part of the dynamics calculation.

POTENTIAL ENERGY FUNCTIONS

For two reasons, the potential energy (PE) plays a central role in molecular dynamics simulation. Firstly, according to Equation (1.3), the force acting on an atom is related to the first derivative of the potential energy function. Secondly, during a molecular dynamics simulation, the total energy (E) must be carefully monitored. This is the sum of the kinetic and potential energies:

$$E = \frac{1}{2} \sum_{i=1}^{N_{df}} m_i v_i^2 + \mathcal{V}(\boldsymbol{r}) \tag{1.12}$$

where N_{df} is the number of degrees of freedom. The kinetic energy is simple to calculate but the PE is a complicated function of atomic positions. This function must be able to predict molecular properties that are measurable by experiment before being applied to situations that are inaccessible to any experimental probes. An excellent review of PE functions has been carried out by Lifson (1981) and recently in several papers (Maple *et al.*, 1988; Hagler *et al.*, 1989; Burt *et al.*, 1989).

The PE function contains many parameters. Ideally, these would be determined by solving the Schrödinger equation, but this is still only possible for a few simple models. The Born–Oppenheimer approximation, which separates the Schrödinger equation into two parts—electronic and nuclear, can be used to obtain a detailed picture of the PE surface for systems of up to 10–20 atoms, but such calculations are not yet feasible for large molecules (McCammon, 1984). Instead an empirical PE function is used, consisting of terms that account for covalent bond stretching, bond angle bending, harmonic dihedral bending, sinusoidal dihedral torsions and non-bonded (van der Waals and Coulombic) interactions. (Quantum corrections to this classical approach have been shown to be very small, being significant for local vibrations with frequencies above 300 cm^{-1} (Karplus and McCammon, 1981).) This energy function can be considered as an approximation of the Born–Oppenheimer type aimed at representing the surface at the second level (i.e. nuclear) (Lifson, 1981). This approximation can be understood as applicable when we consider that during a small displacement of the nucleus the electrons that surround it will have circled it many thousands of times and that therefore their effect can be considered as an average shell. The parameters in empirical PE functions are obtained from experimental and quantum mechanical studies. One computer program that is widely used for molecular dynamics of macromolecules is GROMOS87. The GROMOS87 PE function has the form (Aqvist *et al.*, 1985)

$$\mathcal{V}(\boldsymbol{r}) = \mathcal{V}(\boldsymbol{r}_1, \boldsymbol{r}_2, \ldots, \boldsymbol{r}_{N_{\mathrm{at}}}) \tag{1.13}$$

$$= \underbrace{\sum_{n=1}^{N_b} \frac{1}{2} K_{b_n}(b_n - b_{0_n})^2}_{\text{bonds}} + \underbrace{\sum_{n=1}^{N_\theta} \frac{1}{2} K_{\theta_n}(\theta_n - \theta_{0_n})^2}_{\text{angles}}$$

$$+ \underbrace{\sum_{n=1}^{N_\xi} \frac{1}{2} K_{\xi_n}(\xi_n - \xi_{0_n})^2}_{\text{improper torsions}} + \underbrace{\sum_{n'=1}^{N_\phi} K_{\phi_{n'}}[1 + \cos(n_{n'}\phi_{n'} - \delta_{n'})]}_{\text{proper torsions}}$$

$$+ \sum_{i<j}^{N_{\mathrm{at}}} \left[\underbrace{A_{i,j}/r_{ij}^{12} - C_{i,j}/r_{ij}^{6}}_{\text{van der Waals}} + \underbrace{q_i q_j/(4\pi\varepsilon_0\varepsilon_r r_{ij})}_{\text{electrostatic}} \right]$$

$$+ \text{ additional terms} \tag{1.14}$$

where $\boldsymbol{r}_1, \boldsymbol{r}_2, \ldots, \boldsymbol{r}_{N_{\mathrm{at}}}$ represent the coordinates of the system that is to be investigated. These may be the explicit atomic Cartesian coordinates or the internal coordinates (bond lengths, bond angles and torsional angles) from which all interatomic distances can be calculated. Work is currently in progress to determine better parameters and potential energy functions (Palca, 1986).

Bond Potential

It is possible to obtain a good fit to the second level of the Born–Oppenheimer approximation for bonds using the Morse potential

$$\mathcal{V}_{\text{bonds}} = \sum_{n=1}^{N_b} (D_{b_n}\{1 - \exp[-\alpha_n(b_n - b_{0_n})]\}^2 - D_{b_n}) \tag{1.15}$$

where $-D_{b_n}$ is the energy at the equilibrium bond length b_{0_n} and α_n is used to adjust the potential so that it is able to produce the molecular vibration spectrum of diatomic molecules. For general situations where the fluctuations of bond lengths are close to the equilibrium value, a simple harmonic approximation is sufficient, as in Equation (1.14), where $K_{b_n} = 2D_{b_n}\alpha_n^2$. These two curves are shown in Figure 1.1.

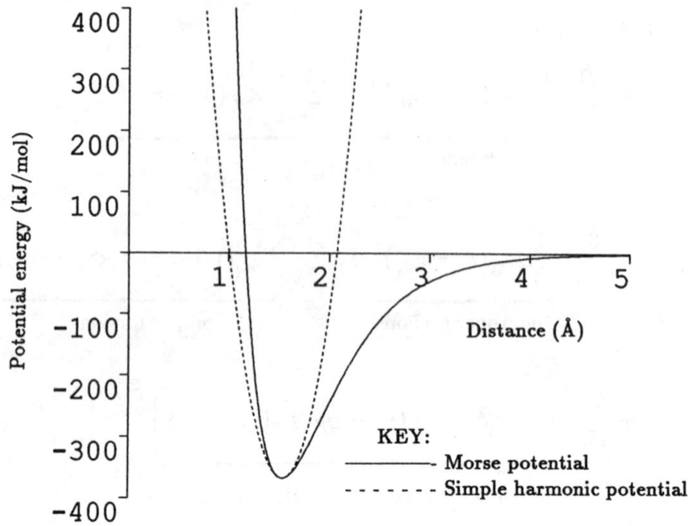

Figure 1.1 Comparison of the Morse and simple harmonic bond potentials where $K_{b_n} = 2D_{b_n}\alpha_n^2$.

Bond Angles

Bond angles are understood principally from a qualitative point of view and therefore the derivation of a PE function describing bond angles is not so simple. For small deviations, treating them as quadratic is a reasonable approximation and, as no better function has been identified, this has become the function of choice for peptides and proteins. This potential is not very accurate for large deviations and this fact should be borne in mind. In Equation (1.14) K_{θ_n}, θ_n and θ_{0_n} correspond to the equivalent bond angle parameters.

Torsions

Proper torsions (or dihedral rotations) describe the rotation around a given bond (see Figure 1.2). The most commonly used form of potential function using torsion angles is known as the Pitzer potential. In Equation (1.14), $K_{\phi_{n'}}$ is the force constant, $\delta_{n'}$ is the reference angle where the potential energy is a maximum and $n_{n'}$ is the multiplicity, that is the number of potential minima in one full rotation. The peptide bond has a partial double bond character with a torsional energy barrier of around 20 kcal mol^{-1}. All peptide bonds are usually found as *trans* in proteins except prolines which occasionally occur as the *cis* isomer. The boundaries

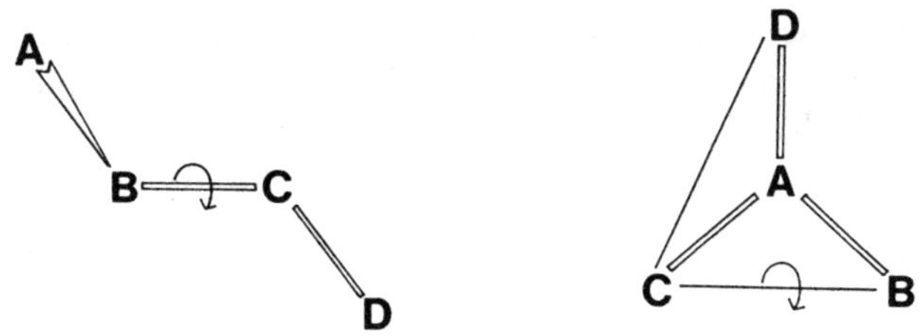

1 2

Figure 1.2 Definitions of (1) proper torsion and (2) improper torsion angles. Open bars show the connecting bonds between atoms A, B, C and D. The thin lines represent connections defined purely for calculation purposes. In both cases the angle is defined as that between the planes ABC and BCD, as shown

for the ϕ, ψ and χ torsional angles are much lower. The nature of torsional potentials is not fully understood and there is even some dispute of the periodicity and location of the energy minima of these functions (Lifson, 1981).

Improper torsions are the out-of-plane bending of a central atom with the three surrounding atoms to which it is bonded (see Figure 1.2). These are particularly important when the united atom approach is used (described later). As with bond angles the theoretical basis is not strong, but deviations from the equilibrium value are small in proteins and the quadratic form is sufficient. In Equation (1.14) K_{ξ_n}, ξ_n and ξ_{0_n} correspond to the equivalent bond parameters.

Cross-terms

Some force fields contain additional potential terms that account for coupling of interactions. These are known as cross-terms and may take the form

$$\begin{aligned}\mathcal{V}_{\text{ct}} = & \sum_{n=1}^{N_{bb'}} \underbrace{F_{bb'}(b_n - b_{0_n})(b'_n - b'_{0_n})}_{\text{bond–bond}} + \sum_{n=1}^{N_{b\theta}} \underbrace{F_{b\theta}(b_n - b_{0_n})(\theta_n - \theta_{0_n})}_{\text{bond–angle}} \\ & + \sum_{n=1}^{N_{\theta\theta'}} \underbrace{F_{\theta\theta'}(\theta_n - \theta_{0_n})(\theta'_n - \theta'_{0_n})}_{\text{angle–angle}} + \sum_{n=1}^{N_{\xi\xi'}} \underbrace{F_{\xi\xi'}(\xi_n - \xi_{0_n})(\xi'_n - \xi'_{0_n})}_{\text{improper torsion–improper torsion}} \\ & + \sum_{n=1}^{N_{\phi\theta\theta'}} \underbrace{F_{\phi\theta\theta'} \cos\phi_n (\theta_n - \theta_{0_n})(\theta'_n - \theta'_{0_n})}_{\text{proper torsion–angle–angle}} \end{aligned} \qquad (1.16)$$

Previously these terms have often been neglected and only included when calculating vibrational spectra. The importance of these terms has been recently investigated and the results would suggest that the use of cross-terms is important for properly evaluating energetics and geometry (Hagler *et al.*, 1989).

Non-bonded Interactions

Non-bonded interactions are split into three parts:

1. dispersion (attractive) term;
2. repulsion term;
3. Coulombic (electrostatic) term.

Calculation of long-range interaction could be carried out by a variety of methods, such as Ewald summation. Unfortunately these are computationally very expensive and generally only applicable to infinitely periodic systems (e.g. crystals). It is not possible to obtain exact solutions to the interactions between two atoms. Generally, information has been obtained by extensive quantum mechanical calculations and crystal packing data. The first two terms account for the van der Waals interaction and all take a similar form. These interactions are not calculated for atoms separated by two consecutive bonds as they would be very large. The non-bonded interactions for these are effectively supplied by the first four terms in Equation (1.14). The interaction with the third neighbour is adjusted so that its contribution to the torsional barrier is taken into account.

Dispersion Terms

At distances where the electron clouds do not overlap, the atoms will experience a net attractive force. This is because at any given moment there are instantaneous dipoles created as a result of the nuclear and electronic fluctuations. Dipoles are induced in other atoms by this, producing a dispersive force. The work in this area is mainly due to that of London where the attractive force has the form of a power series:

$$\mathcal{V}_{\mathrm{disp}} = -\frac{C_{ij}}{r_{ij}^{6}} - \frac{C'_{ij}}{r_{ij}^{8}} - \frac{C''_{ij}}{r_{ij}^{10}} - \cdots \tag{1.17}$$

as $-C_{ij}/r_{ij}^{6}$ is the dominating term at long distances the remainder is neglected. It is important to remember that this is an approximation and that at closer distances the remaining terms may be significant. Slater and Kirkwood derived a theoretical relationship for C_{ij} by finding the best fit using experimental data from work on noble gases. This takes the form

$$C_{ij} = \frac{3e\hbar}{2M_{\mathrm{e}}^{1/2}} \frac{\alpha_i \alpha_j}{(\alpha_i/N_i)^{1/2} + (\alpha_j/N_j)^{1/2}} \tag{1.18}$$

where α and N are the polarizabilities and number of electrons in the outer shell for atoms of the type i and j respectively. As mentioned earlier, this equation is derived from work on the noble gases and has been assumed to be applicable to polyatomic molecules, but never proved. Although there are only a limited set of atoms that occur in macromolecules, we would require $N(1+N)/2$ non-independent parameters, most of which cannot be determined by empirical methods. It is therefore beneficial to determine a combination rule which enables the calculation of these coefficients from those parameters that are obtainable from empirical methods, i.e. from C_{ii}. From Equation (1.18) we note that the coefficients C_{ij} relate to C_{ii} and C_{jj} by the following rule:

$$2\frac{\alpha_i\alpha_j}{C_{ij}} = \frac{\alpha_i^2}{C_{ii}} + \frac{\alpha_j^2}{C_{jj}} \tag{1.19}$$

It is now possible to calculate all of these coefficients from N empirical parameters. This is usually taken one step further using the geometric mean:

$$C_{ij}^2 = C_{ii}C_{jj} \tag{1.20}$$

Although this has the advantage of avoiding the use of polarizabilities it is exceptionally inaccurate. Using Equation (1.19) the RMS deviation from experiment was 3.3% as opposed to 73.5% when using Equation (1.20) (Lifson, 1981).

Repulsive Terms

When atoms come close to each other their electron clouds overlap and produce a repulsive force. This force rapidly increases as the two atoms become closer and the nature of this force is considered to be exponential. In simple terms this may be expressed as

$$\mathcal{V}_{\text{rep}} = A\mathrm{e}^{-br_{ij}} \tag{1.21}$$

Determination of A and b is not a trivial matter and in any case this form of the repulsion term is not computationally efficient. As the nature of this function is to create a steep repulsive force as two atoms become close, this term can be replaced by one relying on high powers of r_{ij}^n. The work in this area was principally carried out by Lennard-Jones, who found that this power was arbitrary as long as it was sufficiently steep. Often $n = 9$ is used but more commonly $n = 12$, as this is computationally more convenient. Once again the geometric mean is used to produce all the coefficients. The resultant potential is that shown in Equation (1.14).

Electrostatic Terms

The elecrostatic interactions are an important part of the force field that is used for macromolecular calculations as these systems contain many highly polar groups. The electrostatic PE function in Equation (1.14) is that given by Coulomb's law, where q_i and q_j are the partial charges and ε_r is the relative dielectric permittivity. This potential can be extended to include higher-order terms that account for dipole moments, quadrupole moments, etc.

The choice of a value for the relative dielectric permittivity has in the past been the cause of some dispute. In general, values of $\varepsilon = 1$–8 (in one case $\varepsilon = 20$; Rees, 1980) have been used depending on the method by which the atomic charges have been calculated (Burt *et al.*, 1989). In some recent work, the use of a distance-dependent dielectric constant has been incorporated (Brooks *et al.*, 1983), thereby damping the long-range interactions more than those at shorter ranges. The meaning of the relative dielectric constant is very complex and it is reasonable to set it to unity for most applications (Lifson, 1981). Partial atomic charges are usually calculated using Mulliken population analysis or derived from fitting to crystal geometries and parameters.

Hydrogen Bonds

The nature of the origin of hydrogen bonds has been a matter of some debate since first being recognized some 80 years ago. As they are one of the most prominent features and strongest non-bonded interactions found in biological systems it is important that any PE function must be able to reproduce them.

In some potential energy functions an explicit term for hydrogen bonding is introduced to obtain reasonable equilibrium hydrogen bond distances (Brooks *et al.*, 1983), especially if the Coulombic term is neglected (Levitt, 1983). In these terms the hydrogen bond is given a distance and angular dependence.

Other groups have not found this extra term to be necessary, accounting for the hydrogen bond as a function of van der Waals and electrostatic interactions. In these cases the usual parameters are used with the exception of the non-bonded parameters of polar hydrogens which are set to zero. The suitability of this assumption has been elegantly demonstrated by Hagler and coworkers (1989) by examining the spatial electron densities.

TREATMENT OF BOUNDARIES

Most of the early simulations were carried out *in vacuo*. That is, the isolated molecule was simulated without any boundary conditions. This makes molecules become distorted, particularly if they are not spherical, as they tend to minimize their surface area. The outermost atoms become unrealistic as they lack any interaction with surrounding atoms. This has a further effect of removing any shielding by the solvent with high dielectric constant on electrostatic interactions. The most recent *in vacuo* simulation has been of the largest protein molecule yet simulated and analysis of the results was limited to the interior of the protein.

To overcome this an effectively infinite system can be produced by introducing periodic boundary conditions. The molecules are confined within a cell which is surrounded by periodic replicas of itself. This means that, whenever a particle leaves the cell through one of its faces, its image in the neighbouring cell on the opposite face takes its place (Figure 1.3). Most simulations employ cubic boundary conditions. However, the most suitable shape, in terms of computational efficiency, is the truncated octahedron (Figure 1.4), which is obtained by symmetrically cutting off the

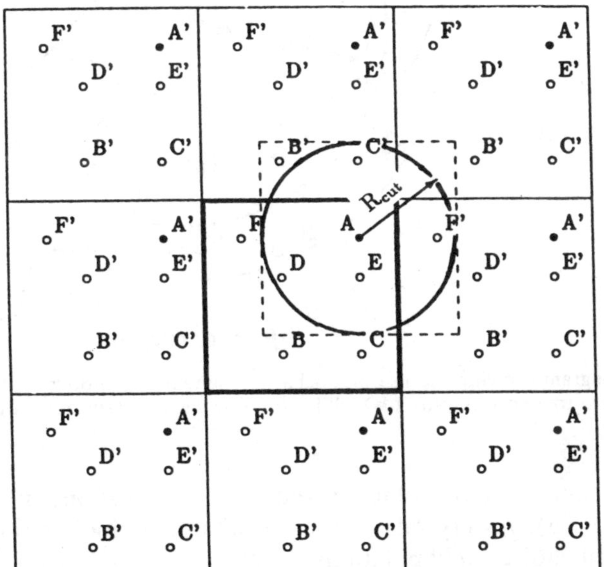

Figure 1.3 Simple periodic boundary conditions. The central box (dark borders) contains the real atoms while the surrounding boxes contain calculated replicas. Atom A interacts with real atoms D and E and with images of C and F, as these lie within the cutoff radius R_{cut}, which must be less than half the box length

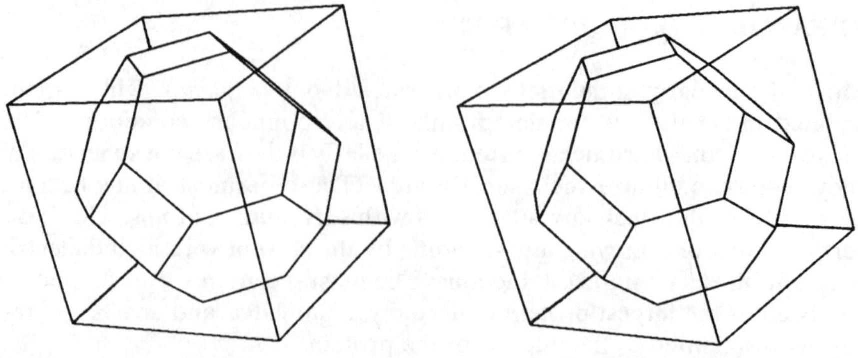

Figure 1.4 Stereo drawing of the truncated octahedron inside the equivalent cube

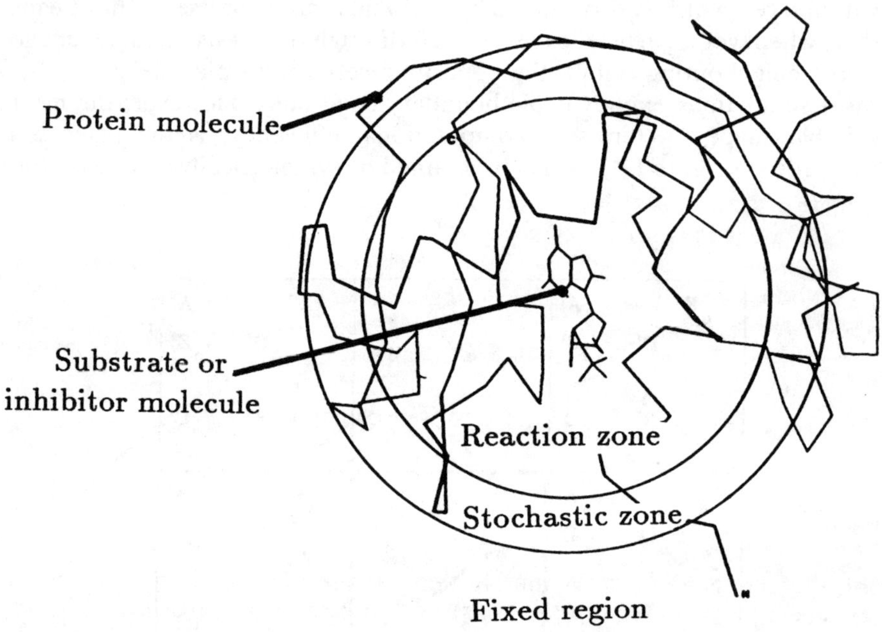

Figure 1.5 Diagram showing the use of stochastic boundary conditions. All atoms in the reaction zone are treated with pure MD while atoms in the stochastic zone are treated with SD

corners of a cube until the volume is reduced by almost one-half (Fincham and Heyes, 1985). As crystals contain a natural periodic structure, they offer a natural choice for simulations.

During simulations of molecules in solvent, particularly with proteins, much of the time is spent simulating bulk water. To reduce this expense, the use of stochastic boundary conditions may be employed. This method can be used to avoid explicit treatment of part of the system whilst maintaining its influence on the remainder. The motion in this region is determined primarily by the time-dependent variations of its non-bonded interactions with neighbouring atoms (Figure 1.5). These interactions produce randomly varying forces which can be approximated by a stochastic force. Stochastic dynamics (SD) is simulated by integrating a set of Langevin equations of motion containing a stochastic R_i and frictional force proportional to a frictional coefficient γ_i:

$$\frac{d^2r_i}{dt^2} = \frac{F_i + R_i}{m_i} - \gamma_i \frac{dr_i}{dt} \tag{1.22}$$

This method has been used to simulate the active site dynamics of ribonuclease (Brünger *et al.*, 1985).

CALCULATION OF NON-BONDED INTERACTIONS

It is computationally far too expensive to calculate non-bonded interactions between all pairs of atoms as the increase in computing time is of the order of N^2. To reduce computing time, a cut off radius R_{cut} is used, outside of which interactions are considered to be negligible and therefore not included. It is generally accepted that R_{cut} should not be lower than 8 Å. As the electrostatic interaction has a $1/r_{ij}$ dependence, its effect at these distances is not negligible. Longer distances of over 15 Å would be required to reduce these errors. To overcome this, local atoms can be treated as charge groups and, if their centroid lies within R_{cut} of the charge group whose interactions are being calculated, then the whole group is considered. If the atomic charges in these groups add up to exactly zero, then the electrostatic interaction of two such groups can be considered to be dipolar in character and to demonstrate a $1/r_{ij}^3$ distance dependence (van Gunsteren and Berendsen, 1987).

As mentioned, calculation of the non-bonded interaction requires a considerable amount of computing time. This requirement can be further reduced by calculating and updating the neighbour list after several steps, typically 10. Furthermore, this list can be split into two parts, one accounting for local interactions and the other for long-range interactions. The long-range interactions list can be updated less frequently. Both of these methods effectively discount the high frequency fluctuations in non-bonded interactions.

A further saving of computing time can be achieved by using a united atom approach for hydrogen atoms that are bonded to carbon atoms. Their influence is maintained by increasing the van der Waals radii of the carbon atoms that they are bonded to. As there are almost as many of these hydrogens as there are other atoms, a significant reduction in computing time is achieved.

CONSTRAINED DYNAMICS

Another method for saving computer time is to constrain the degrees of freedom that have the highest frequency. This then allows a larger integration step Δt to be taken and thereby achieves a simulation of longer length for the equivalent number of iterations. It is important that those degrees of freedom that are removed are only weakly coupled to the remaining ones and are separable from other modes so as not to perturb longer-scale motions.

This principle has been applied successfully to bond lengths, but has been found to be inappropriate for constraining angles from both a physical and a computational point of view. The numerical algorithm has been incorporated into MD simulations as the SHAKE procedure (van Gunsteren and Berendsen, 1977) and allows several-fold increase in step size (up to 2 fs). Effectively an additional zero term is added to the potential that does not contain the term whose force is to be removed. This extra term is the constraint force which compensates for the components of the potential that act along the direction of the constraint.

CONSTANT TEMPERATURE AND PRESSURE

The temperature T is defined by the kinetic energy of a system (Brooks *et al.*, 1983):

$$\sum_{i=1}^{N_{df}} \frac{m_i v_i^2}{2} = \frac{N_{df} K_B T}{2} \tag{1.23}$$

where m_i is the mass of atom i with velocity v_i, N_{df} is the number of degrees of freedom and K_B is Boltzmann's constant. The temperature T can be kept constant during a molecular dynamics simulation by a variety of methods. Probably the most simple method is that derived by Woodcock (Fincham and Heyes, 1985). The temperature at each step is calculated and a position increment scaling factor is determined. This can be incorporated into the Verlet leapfrog procedure. If T_0 is the prespecified temperature and $T(t)$ is calculated from Equation (1.23), then the scaling factor f can be obtained from:

$$f = \sqrt{\frac{T_0}{T(t)}} \tag{1.24}$$

The current atomic positions are then scaled by this factor when predicting the next positions.

Another approach to keeping temperature constant is to couple the system weakly to an external bath (Berendsen *et al.*, 1984; van Gunsteren *et al.*, 1983). This thermal bath is used to scale the velocities at each step by the relation

$$\frac{dT(t)}{dt} = \tau^{-1}[T_0 - T(t)] \tag{1.25}$$

The strength of this coupling is determined by τ, the temperature relaxation time. The velocities are then scaled by a temperature scaling factor:

$$\lambda = \sqrt{1 + \frac{\Delta t}{\tau}\left(\frac{T_0}{T(t)} - 1\right)} \tag{1.26}$$

$$V_i(t) \leftarrow \lambda V_i(t) \tag{1.27}$$

Pressure can also be calculated from MD (van der Ploeg, 1982). This can be kept constant in a similar manner by adjusting the volume of the cell (Fincham and Heyes, 1985; Berendsen *et al.*, 1984).

MOLECULAR DYNAMICS SIMULATIONS

An example of output from a constant-temperature MD simulation of deamino oxytocin is shown in Figure 1.6. In this case, two unit cells of the

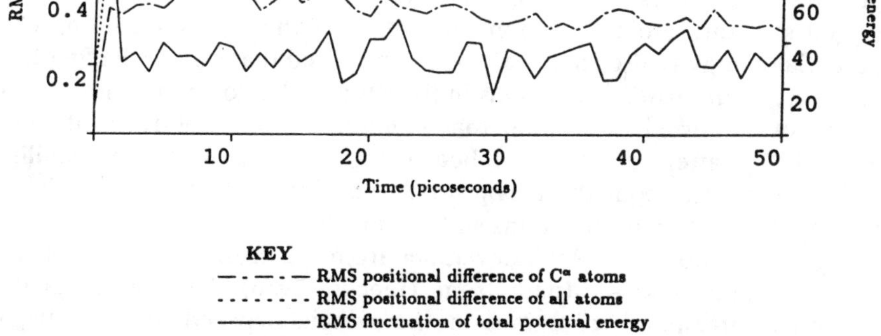

Figure 1.6 Diagram of the RMS fluctuations of potential energy and RMS positional differences between the crystal structure and the MD simulation plotted against simulated time

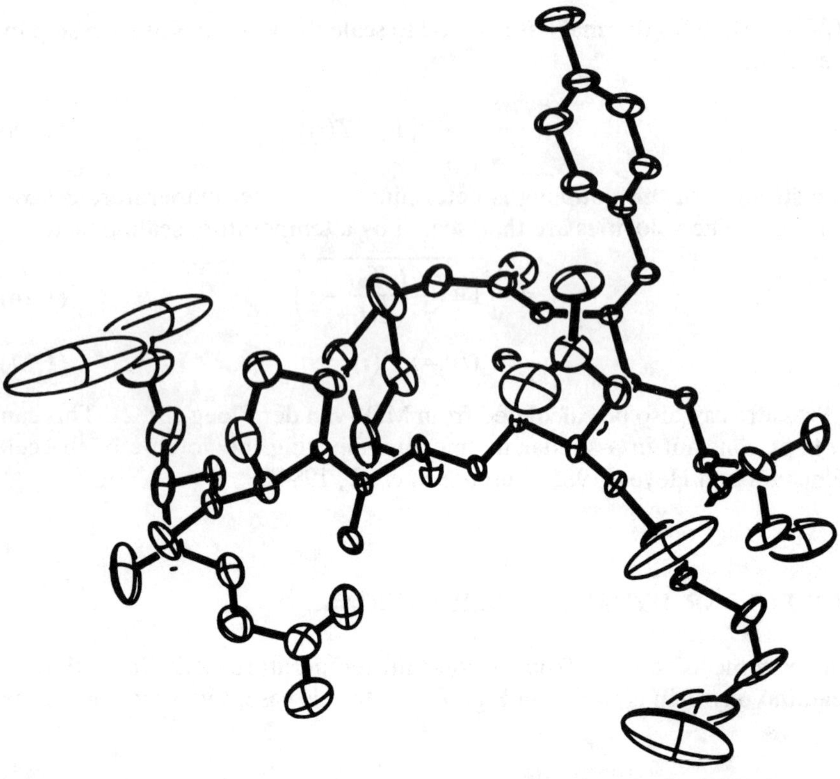

Figure 1.7 Equiprobability ellipsoids drawn about the mean atomic positions

crystal structure were simulated and the structure naturally provided the periodic boundaries. The RMS positional differences are between the crystal structure atomic positions and those of the MD simulation. The potential energy is also shown and it can be noted that during the first few picoseconds the RMS fluctuations in the PE and the positional differences have not stabilized, indicating that equilibrium was not reached until around 10 ps after the simulation began. Figure 1.7 shows equiprobability ellipsoids drawn about the mean atomic positions. Their size and shape indicate the motion of the atoms and its anisotropy.

Figure 1.8 shows the RMS deviations from the crystal structure of the enzyme ribonuclease-A (data from Haneef, 1986). The larger peaks generally correspond to regions of the protein exposed to the solvent where the X-ray structure cannot model the full flexibility of the protein. Further applications of molecular dynamics are described in detail in Chapters 2 and 3 for proteins and carbohydrates respectively.

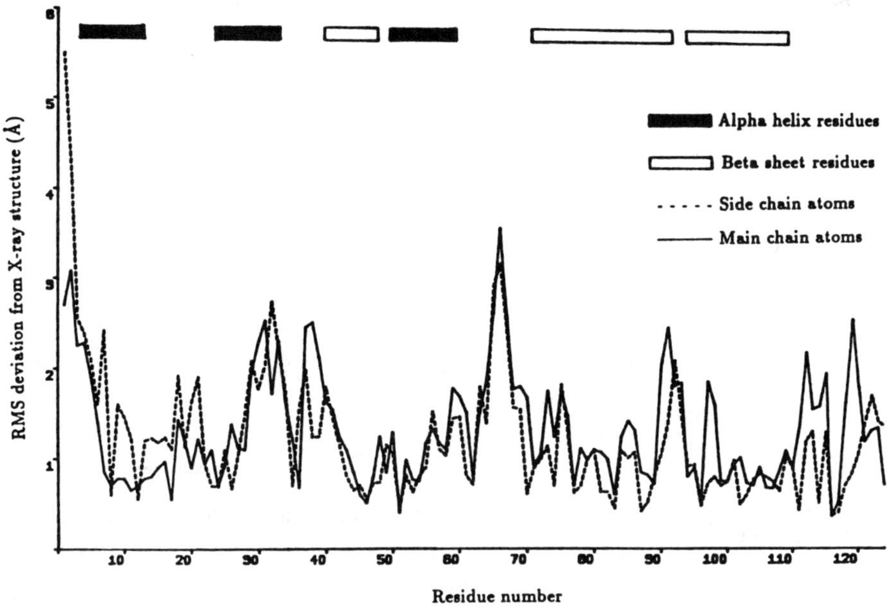

Figure 1.8 RMS deviations of the average MD structure from the crystal structure

SOLUTION OF NMR STRUCTURES

With the advent of two-dimensional nuclear magnetic resonance (NMR), significant progress has been made towards the determination of macromolecules with molecular weights of up to 10 000 daltons (Wagner and Wüthrich, 1982). This has meant that the three-dimensional structure of macromolecules can be elucidated in different solutions without the need to crystallize them first. By identification of nuclear proton–proton distances from nuclear enhancement spectra, it is possible to predict the three-dimensional structure by distance geometry. Such a procedure has been applied to lipid-bound glucagon (Braun *et al.*, 1983), proteinase IIA (Williamson *et al.*, 1985) and bovine pancreatic trypsin inhibitor (Havel and Wüthrich, 1985). To improve the quality of the final structures further, the interproton distances can be incorporated into molecular dynamics simulations by using a pseudo-potential term introduced into the potential energy function. This additional term takes the form of a skewed biharmonic effective potential (Clore *et al.*, 1985; Figure 1.9). This term is incorporated as a restraint term:

$$\mathcal{V}_{\rm res}(r_{ij}) = \begin{cases} \frac{1}{2}K^{\rm u}_{\rm res}(r_{ij}-r^{\rm u}_{ij})^2 & \text{if } r_{ij}>r^{\rm u}_{ij} \\ 0 & \text{if } r^{\rm l}_{ij}\le r_{ij}\le r^{\rm u}_{ij} \\ \frac{1}{2}K^{\rm l}_{\rm res}(r_{ij}-r^{\rm l}_{ij})^2 & \text{if } r_{ij}<r^{\rm l}_{ij} \end{cases} \tag{1.28}$$

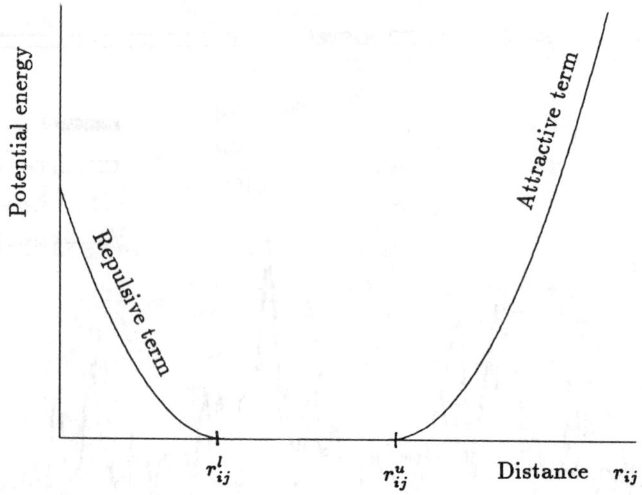

Figure 1.9 Graphical representation of the skewed biharmonic effective potential used as an NOE restraint

where r_{ij}^{u} and r_{ij}^{l} are the upper and lower distances respectively. $K_{\mathrm{res}}^{\mathrm{u}}$ and K_{res}^{l} are upper and lower restraint force constants respectively. The magnitude of this parameter must be chosen carefully so that the restraining term is balanced and will lead to the best sampling of conformations. The value can be coupled to the error within the NMR experiments as

$$K_{\mathrm{res}} = \frac{k_{\mathrm{B}}TS}{2(\Delta_{ij})^2} \qquad (1.29)$$

where S is a scaling factor and Δ_{ij} are the positive and negative error estimates of r_{ij} for the upper and lower values respectively. The use of this restraining term is not restricted to NMR. Any modelling that requires restraining of atomic distances can utilize this force.

As the only possible conformations that satisfy a set of given restraints are being searched for, it is not necessary to enforce the whole force field exactly. In fact, it is often advantageous to alter many of the parameters, in particular non-bonded interactions, in order to allow the atoms to pass each other. After this the full force field is slowly built up again. When using a united atom approach, it is necessary to use pseudo-atoms to represent those hydrogens that are not included in the calculation, but are identified to be within nuclear Overhauser effect (NOE) distances found by NMR experiments. A detailed account of the applications of molecular dynamics in the solution of NMR structures is given in Chapter 5.

REFINEMENT OF X-RAY CRYSTALLOGRAPHIC DATA WITH MOLECULAR DYNAMICS

In a similar way to NMR refinement, it is possible to include an energy function that can aid the refinement of X-ray structures (Fujinaga *et al.*, 1989; Brünger, 1988). The aim is to minimize the difference between calculated X-ray diffraction structure factor amplitudes ($|F_{calc}(h,k,l)|$), where h, k and l are the reciprocal lattice points of the crystal) and the observed ones ($|F_{obs}(h,k,l)|$), where the structure amplitude is related to the X-ray reflection intensities: $|F(h,k,l)|^2 \propto I(h,k,l)$. When the sum of weighted residuals is a minimum, the quality of this difference is generally quoted by its R factor:

$$R = \frac{\sum_{h,k,l} \| F_{obs}(h,k,l)| - |F_{calc}(h,k,l)\|}{\sum_{h,k,l} |F_{obs}(h,k,l)|} \tag{1.30}$$

The structure factor is defined as the Fourier transform of the electron density, $\rho(x,y,z)$:

$$F(h,k,l) = \int\int\int \rho(x,y,z) e^{2\pi i(hx+ky+lz)} \, dx \, dy \, dz \tag{1.31}$$

Traditionally a least-squares refinement method is employed to minimize this difference. However, this procedure becomes trapped in local minima, requiring human intervention to rebuild regions in the current model which poorly fit the observed electron density. To increase the radius of convergence of the refinement, the following PE term can be included in the standard empirical PE function:

$$\mathcal{V}_{X\text{-ray}} = K_{X\text{-ray}} \sum_{h,k,l} [|F_{obs}(h,k,l)| - |F_{calc}(h,k,l)|]^2 \tag{1.32}$$

In a similar way to NMR it is recommended that initial calculations start with low resolution intensities which slowly increase during the simulation. Applications of this method are described in Chapter 5.

CALCULATION OF FREE ENERGIES OF PERTURBATION

In recent years, calculation of free energies of perturbation has become one of the most important applications of MD simulation (Beveridge and DiCapua, 1989; van Gunsteren, 1989; Singh *et al.*, 1987; Warshel *et al.*, 1988; Chapter 5). Free energy differences are related to the equilibrium constant K as previously described in Equation (1.1) and hence protein–ligand binding equilibria can be studied by MD simulations. The most commonly used approach in conjunction with MD is that of thermodynamic integration, whereby the difference in free energy of two states is calculated. This can be achieved either by simulating a number of intermediate steps between the two states, or by slowly changing the parameters of one state into the other state. These methods are called **windowing** and **slow growth** respectively, and are described below.

If we have two physical states, a and b, that are slightly different in character, they will have a difference in free energy ΔG. Between these two states we can consider many intermediate states that are separated by very small spaces λ, the sum of which adds up to the free energy difference between a and b, that is

$$\Delta G = G_{\mathrm{b}} - G_{\mathrm{a}} = \sum_{i=1}^{N_{\mathrm{S}}} \Delta G_i \tag{1.33}$$

From basic statistical mechanics it can be shown that

$$\Delta G_i = -RT \ln \left\langle \exp\left[-\frac{\mathcal{V}(\lambda_i) - \mathcal{V}(\lambda_{i-1})}{RT} \right] \right\rangle_{\lambda_{i-1}} \tag{1.34}$$

where $\langle\rangle$ denotes the average of the equilibrium ensemble of states corresponding to λ_{i-1}. At each value λ_{i-1} an MD simulation is carried out and the average in Equation (1.34) is evaluated. λ is altered in steps from zero to one by spacings of $\delta\lambda_i$. These spacings do not need to be equal and it may be more efficient to vary this value. This method is known as windowing.

In the slow growth method, $\delta\lambda_i$ is taken as exceedingly small so that we can equate ΔG_i with a single instantaneous value:

$$\Delta G_i \approx -RT \ln \exp\left[-\frac{\mathcal{V}(\lambda_i) - \mathcal{V}(\lambda_{i-1})}{RT} \right] \tag{1.35}$$

$$= \mathcal{V}(\lambda_i) - \mathcal{V}(\lambda_{i-1}) \tag{1.36}$$

and therefore the overall difference in free energy can be expressed:

$$\Delta G = \sum_{i=1}^{N_{\mathrm{s}}} \mathcal{V}(\lambda_i) - \mathcal{V}(\lambda_{i-1}) \tag{1.37}$$

This is easily incorporated into the PE function where each term is coupled to the spacing parameter λ. For instance, for bonds the potential term is

$$\mathcal{V}_{\mathrm{bonds}} = \sum_{n=1}^{N_{\mathrm{b}}} \frac{1}{2} [(1-\lambda)K^{\mathrm{a}}_{b_n} + \lambda K^{\mathrm{b}}_{b_n}][b_n - ((1-\lambda)b^{\mathrm{a}}_{0_n} + \lambda b^{\mathrm{b}}_{0_n})]^2 \tag{1.38}$$

Setting $\lambda = 0$, Equation (1.38) gives the contribution to the PE for a bond of equilibrium length $b^{\mathrm{a}}_{0_n}$ and force constant $K^{\mathrm{a}}_{b_n}$. When $\lambda = 1$, the equation gives the PE for a bond with parameters $b^{\mathrm{b}}_{0_n}$ and $K^{\mathrm{b}}_{b_n}$. By carrying out the MD while gradually varying λ and evaluating the equation (1.36) at each step, the free energy difference between the system with bond a and that with bond b can be evaluated. The fact that intermediate values of λ do not correspond to chemically realistic molecules is irrelevant, because ΔG is independent of the path that is taken between the two states.

In practice, this procedure is usually used to evaluate ΔG values in a thermodynamic cycle and yields the changes in the free energy of binding when engineered changes to a protein or ligand are carried out. This method may be extended to the calculation of the changes in enzyme rate constants by using MD simulations of transition state complexes.

OTHER METHODS EMPLOYED IN MOLECULAR SIMULATIONS

In addition to molecular dynamics there are other techniques which enable the calculation of structural and thermodynamic properties of biological molecules. These methods include

1. energy minimization (molecular mechanics calculations);
2. normal mode analysis;
3. Monte Carlo simulations.

Molecular simulation begins with a low energy conformation of the macromolecule in its aqueous environment. Because of various errors even an X-ray structure may have a high energy and energy minimization must be carried out before simulation begins.

Energy Minimization

The structure of minimum potential energy is one whose derivatives of $\mathcal{V}$ are zero with respect to $\boldsymbol{r}$. This is more commonly known as the energy minimum and can be expressed as when $\partial\mathcal{V}(\boldsymbol{r})/\partial r_i$ for all coordinates, i, is zero. This first derivative contains information about the local slope and therefore one method in which to find this equilibrium structure is to move numerically in the direction of the minima, i.e. down the gradient. This simplest of methods is known as steepest descent, each step of which can be written as

$$\boldsymbol{r}_{i+1} = \boldsymbol{r}_i - S_i \frac{\partial\mathcal{V}}{\partial \boldsymbol{r}} \tag{1.39}$$

The step size can be increased if the total energy decreases in the step, i, as $S_{i+1} = 2S_i$ or decreased if the energy increases: $S_{i+1} = S_i/2$. This process is stopped when the difference between the new and the old PEs is lower than a set threshold or, alternatively, after a preset number of iterations.

Other adaptions of this method find the minima along this line by calculating several points along the direction of the gradient and interpolating them to find the minimum point along this direction. Both methods are exceedingly inefficient especially when there are long narrow valleys, even if the valley is perfectly quadratic. The new gradient at the minimum point will be a right angle to the last direction and in general will not be the direction of the minima.

To improve on the shortcomings of steepest descent, a method is required so that a new direction can be constructed to conjugate with all previous directions. Such methods have been determined for quadratic surfaces and are termed conjugate gradient methods, one of which is that due to Fletcher and Reeves. The first step is to carry out a line minimization in the direction of the current gradient (as above). Then the next direction is chosen as a linear combination of the present gradient and the past search direction so that the new direction is linearly independent of all previous directions. For a quadratic function with N variables this method will find the minima in N iterations. The PE function used for biomolecules is not particularly quadratic so after a prespecified number of steps the procedure is restarted from the current points.

Higher-order methods may be used to determine the minima such as the Newton–Raphson method. This method requires a large amount of computer memory and under certain conditions this method is unstable. More importantly, there is not an overwhelming advantage of this method over that of conjugate gradients, which requires much less memory (Press *et al.*, 1988). An application of energy minimization methods is described in Chapter 4.

Normal Mode Analysis

As mentioned in the previous section, the third term contains information about the local curvature of a function. To put it another way, the second derivative of the PE function represents the energy of the system due to small deviations from the equilibrium value. The normal modes of vibration can be calculated from this term if we assume that the system undergoes harmonic motion about its equilibrium configuration. This is equivalent to approximating the potential surface as a multidimensional parabola. At equilibrium the second term in the equation is zero and therefore we can rewrite this equation as

$$\mathcal{V}(\boldsymbol{r}_0+\delta\boldsymbol{r})-\mathcal{V}(\boldsymbol{r}_0)=\tfrac{1}{2}\delta\boldsymbol{r}^{\mathrm{T}}\boldsymbol{F}\,\delta\boldsymbol{r}+\ldots \qquad (1.40)$$

where $\delta\boldsymbol{r}$ is a column vector of displacement coordinates from the equilibrium configuration and $\boldsymbol{F}$ is the matrix of second derivatives of $\mathcal{V}(\boldsymbol{r}_0)$:

$$F = \begin{bmatrix} \frac{\partial^2 \mathcal{V}}{\partial r_1 \partial r_1} & \frac{\partial^2 \mathcal{V}}{\partial r_1 \partial r_2} & \cdots & \frac{\partial^2 \mathcal{V}}{\partial r_1 \partial r_{3N_{at}}} \\ \frac{\partial^2 \mathcal{V}}{\partial r_2 \partial r_1} & \frac{\partial^2 \mathcal{V}}{\partial r_2 \partial r_2} & \cdots & \frac{\partial^2 \mathcal{V}}{\partial r_2 \partial r_{3N_{at}}} \\ \vdots & \vdots & & \vdots \\ \frac{\partial^2 \mathcal{V}}{\partial r_{3N_{at}} \partial r_1} & \frac{\partial^2 \mathcal{V}}{\partial r_{3N_{at}} \partial r_2} & \cdots & \frac{\partial^2 \mathcal{V}}{\partial r_{3N_{at}} \partial r_{3N_{at}}} \end{bmatrix} \quad (1.41)$$

It is better to express $\boldsymbol{r}$ in terms of Cartesian coordinates so that the kinetic energy can take the simple form:

$$T(\delta\dot{\boldsymbol{r}}) = \tfrac{1}{2}\delta\dot{\boldsymbol{r}}^{\mathrm{T}}\boldsymbol{M}\,\delta\dot{\boldsymbol{r}} \quad (1.42)$$

where $\delta\dot{\boldsymbol{r}}$ is the vector of atomic velocities, $\delta\dot{\boldsymbol{r}} = \mathrm{d}\delta\boldsymbol{r}/\mathrm{d}t$, and $\boldsymbol{M}$ is a diagonal matrix of atomic masses of each coordinate. When Cartesian coordinates are used these masses are repeated three times, $\boldsymbol{M} = \mathrm{diag}(m_1, m_1, m_1, m_2, m_2, \ldots, m_{N_{at}})$. Together the potential and kinetic energies form the simultaneous equations of motion:

$$\boldsymbol{M}\,\delta\ddot{\boldsymbol{r}} + \boldsymbol{F}\,\delta\boldsymbol{r} = \boldsymbol{0} \quad (1.43)$$

Here $\delta\ddot{\boldsymbol{r}}$ is the vector of atomic acceleration. Using the simple harmonic approximation we can analytically express the deviations and accelerations of atoms in the following terms:

$$\delta\boldsymbol{r} = \delta\boldsymbol{r}_0\,\mathrm{e}^{2\pi\mathrm{i}\nu t} \quad (1.44)$$

$$\delta\ddot{\boldsymbol{r}} = -(2\pi\nu)^2\,\delta\boldsymbol{r}_0\,\mathrm{e}^{2\pi\mathrm{i}\nu t} \quad (1.45)$$

where ν is the frequency and $\delta\boldsymbol{r}_0$ is the amplitude of vibration. Inserting into Equation (1.43) we obtain:

$$[\boldsymbol{F} - (2\pi\nu)^2\boldsymbol{M}]\delta\boldsymbol{r}_0 = \boldsymbol{0} \quad (1.46)$$

This is then solved for its eigensolutions, where the amplitude of each normal mode is given by the eigenvectors and its frequency by the eigenvalue. There are $3N_{at}$ solutions if this analysis is carried out in Cartesian space, with six of the solutions being zero, representing three translations and three rotations of the molecule. Although this method allows for the analytical determination of molecular motion the use of this method very much depends on how harmonic the motions are around one given protein conformation.

Monte Carlo Simulation

Monte Carlo techniques take a random walk on the potential energy surface, sampling an ensemble of configurations. This method has the advantage that it only requires the PE function in the calculation and not its derivatives. The most commonly used method is that due to Metropolis, where sampling is considered as a Markov process.

From a given starting configuration of PE, $\mathcal{V}_i$, a randomly selected atom is chosen and moved in a random direction by a random amount. The total PE, $\mathcal{V}_j$, of this new configuration is calculated and compared with the previous energy. If $\mathcal{V}_j < \mathcal{V}_i$ then the move is accepted. If, on the other hand, $\mathcal{V}_j > \mathcal{V}_i$ the probability of this occurring at a given temperature T is computed by

$$p_j = \exp\left(-\frac{\mathcal{V}_j - \mathcal{V}_i}{k_B T}\right) \tag{1.47}$$

This probability is then compared with a random number that lies between zero and one. If the value of this random number is lower than p_j then the configuration is accepted. This process is repeated many times to obtain an ensemble of configurations. Unlike molecular dynamics, which calculates atomic trajectories deterministically, the moves calculated by Monte Carlo methods cannot be compared with time-dependent actual motions. Several adjustments can be made to the method. For instance, atoms which lie closer to the region of interest can be sampled preferentially to those that lie further away. In this way more configurational space can be covered for the equivalent amount of computing time.

REFERENCES

Aqvist, J., van Gunsteren, W. F., Leijonmarck, M. and Tapia, O. (1985). A molecular dynamics study of the C-terminal fragment of the L7/L12 ribosomal protein: secondary structure motion in a 150 ps trajectory. *J. Mol. Biol.*, **183**, 461–477

Arseniev, S. A., Kondakov, V. I., Maiorov, V. N. and Brystrov, V. F. (1984). *FEBS Lett.*, **165**, 57–62

Berendsen, H. J. C., Postma, J. P. M., van Gunsteren, W. F., DiNola, A. and Haak, J. R. (1984). Molecular dynamics with coupling to an external bath. *J. Chem. Phys.*, **81**, 3684–3690

Beveridge, D. L. and DiCapua, F. M. (1989). Free energy via molecular simulation: a primer. In van Gunsteren, W. F. and Weiner, P. K. (Eds.), *Computer Simulation of Biomolecular Systems. Theoretical and Experimental Applications*. Escom, Leiden, pp. 1–27

Braun, W., Wider, G., Lee, K. M. and Wuthrich, K. (1983). Conformation of glucagon in a lipid–water interphase by ^{1}H nuclear magnetic resonance. *J. Mol. Biol.*, **169**, 921–948

Brooks, B. R., Bruccoleri, R. E., Olafson, B. D., States, D. J., Swaminathan, S. and Karplus, M. (1983). CHARMM: a program for macromolecular energy minimization, and dynamics calculation. *J. Comput. Chem.*, **4**, 187–217

Brünger, A. T. (1988). Crystallographic refinement by simulated annealing: application to a 2.8 Å resolution structure of aspartate aminotransferase. *J. Mol. Biol.*, **203**, 803–816

Brünger, A. T., Brooks III, C. L. and Karplus, M. (1985). Active site dynamics of ribonuclease. *Proc. Natl. Acad. Sci. USA*, **82**, 8458–8462
Burt, S. K., Mackay, D. and Hagler, A. T. (1989). Theoretical aspects of drug design: molecular mechanics and molecular dynamics. In Penn, T. J. (Ed.), *Computer-Aided Drug Design. Methods and Applications*. Dekker, New York, pp. 55–91
Clore, G. M., Gronenborn, A. M., Brünger, A. T. and Karplus, M. (1985). Solution conformation of a heptadecapeptide comprising the DNA binding helix F of the cyclic AMP receptor protein of *Escherichia coli*. *J. Mol. Biol.*, **186**, 435–455
Fincham, D. and Heyes, D. M. (1985). Recent advances in molecular dynamics computer simulation. *Adv. Chem. Phys.*, **63**, 493–575
Fujinaga, M., Gros, P. and van Gunsteren, W. F. (1989). Testing the method of crystallographic refinement using molecular dynamics. *J. Appl. Crystallogr.*, **22**, 1–8
van Gunsteren, W. F. (1989). Methods for calculation of free energies and binding constants: successes and problems. In van Gunsteren, W. F. and Weiner, P. K. (Eds.), *Computer Simulation of Biomolecular Systems. Theoretical and Experimental Applications*. Escom, Leiden, pp. 27–59
van Gunsteren, W. F. and Berendsen, H. J. C. (1977). Algorithms for macromolecular dynamics and constraint dynamics. *Mol. Phys.*, **34**, 1311–1327
van Gunsteren, W. F. and Berendsen, H. J. C. (1987). GROMOS87. *Groningen Molecular Simulation (GROMOS) Library Manual*. BIOMOS, Nijenborgh 16, Groningen, pp. 1–229
van Gunsteren, W. F., Berendsen, H. J. C., Hermans, J., Hol, W. G. J. and Postma, J. P. M. (1983). Computer simulation of the dynamics of hydrated protein crystals and its comparison with X-ray data. *Proc. Natl. Acad. Sci. USA*, **80**, 4315–4319
Hagler, A. T., Maple, J. R., Thacher, T. S., Fitzgerald, G. B. and Dinur, U. (1989). Potential energy functions for organic and biomolecular systems. In van Gunsteren, W. F. and Weiner, P. K. (Eds.), *Computer Simulation of Biomolecular Systems. Theoretical and Experimental Applications*. Escom, Leiden, pp. 149–167
Haneef, I. (1986). PhD Thesis, London
Havel, T. F. and Wuthrich, K. (1985). An evaluation of the combined use of nuclear magnetic resonance and distance geometry for the determination of protein conformations in solution. *J. Mol. Biol.*, **182**, 281–294
Karplus, M. and McCammon, J. A. (1981). The internal dynamics of globular proteins. *CRC Crit. Rev. Biochem.*, **9**, 293–349
Karplus, M. and McCammon, J. A. (1986). The dynamics of proteins. *Sci. Am.*, **254**, April, 42–51
Kuki, A. and Wolynes, P. G. (1987). Electron tunneling paths in proteins. *Science*, **236**, 1647–1652
Levitt, M. (1983). Molecular dynamics of native protein: II. Analysis and nature of motion. *J. Mol. Biol.*, **168**, 595–620
Lifson, S. (1981). Potential energy functions for structural molecular biology. In *NATO Advanced Study Institute/FEBS Advanced Course No. 78. Current Methods in Structural Molecular Biology*. pp. 1–27
Maple, J. R., Dinur, U. and Hagler, A. T. (1988). Derivation of force fields for molecular mechanics and dynamics from *ab initio* energy surfaces. *Proc. Natl. Acad. Sci. USA*, **85**, 5350–5354
McCammon, J. A. (1984). Protein dynamics. *Rep. Prog. Phys.*, **47**, 1–46
Palca, J. (1986). Cooperation on new molecules. *Nature*, **322**, 586
van der Ploeg, P. (1982). PhD Thesis, Groningen
Press, W. H., Flanny, B. P., Teukolskey, S. A. and Vetterling, W. T. (1988). *Numerical Recipes: the Art of Scientific Computing*. Cambridge University Press, Cambridge, pp. 307–330
Rees, D. C. (1980). Experimental evaluation of the effective dielectric constant of proteins. *J. Mol. Biol.*, **141**, 323–326
Singh, U. C., Brown, F. K., Bash, P. A. and Kollman, P. A. (1987). An approach to the application of free energy perturbation methods using molecular dynamics. *J. Am. Chem. Soc.*, **109**, 1607–1614
Verlet, L. (1967). Computer 'experiments' on the classical fluids. I. Thermodynamical properties of Lennard-Jones molecules. *Phys. Rev.*, **159**, 98–103

Wagner, G. and Wüthrich, K. (1982). Amide proton exchange and surface conformation of the basic pancreatic trypsin inhibitor in solution. Studies with two-dimensional nuclear magnetic resonance. *J. Mol. Biol.*, **160**, 343–361

Warshel, A., Sussman, F. and Hwang, J.-K. (1988). Evaluation of catalytic free energies in genetically modified proteins. *J. Mol. Biol.*, **201**, 139–159

Williamson, M. P., Havel, T. F. and Wuthrich, K. (1985). Solution conformation of proteinase inhibitor IIA from bull seminal plasma by ^{1}H nuclear magnetic resonance and distance geometry. *J. Mol. Biol.*, **182**, 295–315

Zheng, C., Wong, C. F., McCammon, J. A. and Wolynes, P. G. (1988). Quantum simulation of ferrocytochrome c. *Nature*, **334**, 726–728

2
Conformational Variability of Insulin: a Molecular Dynamics Analysis

Leo S. D. Caves, Dzung T. Nguyen and Roderick E. Hubbard

INTRODUCTION

The three-dimensional atomic structure of insulin continues to be the focus of intense study. Since the first description of the insulin structure 20 years ago (Adams *et al.*, 1969), X-ray crystallographic analysis has provided many structures of the native molecule at high resolution (Derewenda *et al.*, 1989), and has revealed that the insulin molecule can adopt a number of similar, but distinct conformations (Dodson *et al.*, 1980a, 1980b; Cutfield *et al.*, 1981). The insulin molecule exhibits both small- and large-scale conformational changes. These have been observed both in the crystal, via X-ray diffraction (Bentley *et al.*, 1978) and thermal diffuse scattering (TDS) (Caspar *et al.*, 1988) experiments, and in solution via circular dichroism (CD) (Wollmer *et al.*, 1977; Renscheidt *et al.*, 1984) and nuclear magnetic resonance (NMR) (Ramesh and Bradbury, 1986; Kaarlsholm *et al.*, 1989; Weiss *et al.*, 1989) experiments.

One important aim of these structural studies has been to establish the relationship between insulin structure and function (Blundell *et al.*, 1972; Baker *et al.*, 1988). Recent studies of modified insulins have indicated that the potential for conformational change may be important for the biological activity of the molecule (Dodson *et al.*, 1983; Baker *et al.*, 1988). Thus the observed conformational states of insulin may be relevant to insulin action and are important models of the types and mechanisms of conformational change in proteins (Chothia *et al.*, 1983).

Aspects of the conformational variability of a protein molecule can be modelled using the molecular dynamics (MD) simulation method (Brooks *et al.*, 1988; McCammon and Harvey, 1987). In a molecular dynamics simulation the detailed time evolution of a system of particles may be evaluated by integrating equations of motion for the particles on their potential energy surface (Alder and Wainright, 1959). For proteins, the molecular potential energy is described by an empirical interaction function which is typically a discrete sum of terms representing bonded and non-bonded atomic interactions (McCammon and Harvey, 1987). The form and parameters of such functions are approximate and have some major limitations (Hermans, 1985), but have provided an effective model of protein interactions at a reasonable computing cost.

From the first protein molecular dynamics simulation (McCammon *et al.*, 1977), the MD method has provided detailed insight into the nature, magnitude and distribution of many aspects of protein motion on the picosecond to nanosecond (10^{-12}–10^{-9} s) timescale (see for example Swaminathan *et al.*, 1982; Åqvist *et al.*, 1985). The MD method has been used to investigate the nature of observable processes that are not understandable from a static picture of protein structure such as tyrosine ring-flipping in the interior of the bovine pancreatic trypsin inhibitor (McCammon *et al.*, 1983) and ligand binding to haem proteins (Case and Karplus, 1979). The method has also been used to address the effect of solvation on protein and solvent structure and dynamics (Brooks and Karplus, 1986, 1989). Through the use of statistical mechanical perturbation theory, the MD method has been used to investigate the thermodynamics of solvation (Bash *et al.*, 1987b) and conformational equilibria (Anderson and Hermans, 1988) in peptides, substrate binding in enzymes (Wong and McCammon, 1986) and simple chemical reactions in solution (Bash *et al.*, 1987a). The MD method has also found powerful application in the interpretation of experimental data from nuclear magnetic resonance (NMR) (Kaptein *et al.*, 1985) and X-ray diffraction (Brünger *et al.*, 1987) experiments.

The work described here focuses on a molecular dynamics analysis of the structures of pig insulin as observed in a number of different crystal forms. From these structures different conformational states of the molecule can be distinguished. Some of the conformers from different crystals are found to be very similar; others exhibit large differences (Cutfield *et al.*, 1981). The differences between conformers have been attributed to the effects of crystal packing (Baker *et al.*, 1988) or medium effects (Bentley *et al.*, 1978). In the molecular dynamics simulations the conformers are released from their crystal environment and their intrinsic (vacuum) potential energy surfaces are sampled with a kinetic energy corresponding to room temperature. This raises a number of interesting questions. How much do the conformers vary from their initial conformation—are some conformers more stable than others? How similar are the dynamic properties of the different conformers? Do the differences between conformers diminish as a result of simulation? How do the simulated dynamics relate to the conformational variability observed in the crystal?

In order to address the above questions, an analysis of simulated trajectories of the different insulin conformers has been performed. The analysis includes an examination of atomic and secondary structure motion and a study of the relationship of time-averaged and transient dynamic structures. The results are related, wherever possible, to data derived from X-ray crystallography. Before the analysis of the trajectories is described, there is a brief description of the insulin structures used and how the molecular dynamics simulations were performed. A preliminary account of this study has been presented (Caves, 1989a).

MATERIALS AND METHODS

Insulin Structure

Chemical Structure and Assembly

Insulin consists of two polypeptide chains, the A-chain of 21 residues and the B-chain of 30 residues (the amino acid sequence of pig insulin is given in Figure 2.1). The chains are interconnected by two disulphide bridges (between A7 and B7, A20 and B19) and there is a further disulphide bridge within the A-chain (between A6 and A11).

Insulin has a strong tendency to self-associate; it exists as a monomer only at very low concentration ($<10^{-7}$ molar) at neutral pH (Frank *et al.*, 1972). At higher concentrations it forms dimers, and in the presence of zinc (or other divalent metal) ions it further assembles to form a hexamer (Goldman and Carpenter, 1976). This facile aggregation into hexamers is thought to be important in the biosynthesis, processing and storage of insulin (Blundell *et al.*, 1972). The basic three-dimensional arrangement of the molecule as revealed by X-ray analysis is now described.

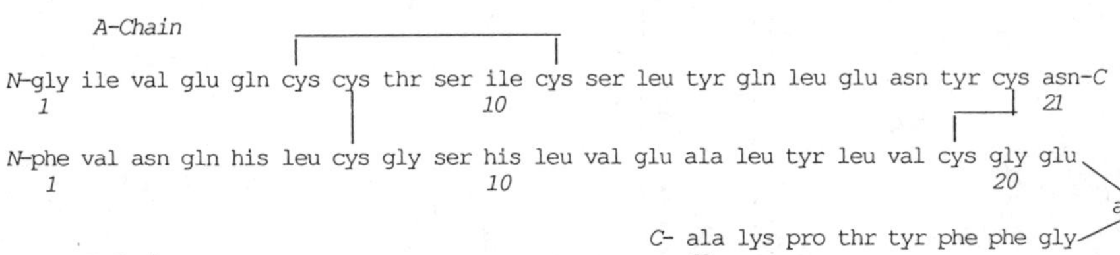

Figure 2.1 The amino acid sequence of pig insulin

Three-dimensional Structure

X-ray diffraction analysis of the native insulin molecule has been performed on different crystal forms and on crystals obtained from solutions of differing composition (Derewenda *et al.*, 1989). The structures (in a

Table 2.1 Crystal structures of pig insulin used in this study

Code	*Crystal system*	*Level of assembly*	*Asymmetric unit*	*Space group*	*Resolution* (Å)	*R-factor* (%)	*Solvent content* (%)	*Comment*	*References*
2ZN	Rhombohedral	Hexamer	Dimer	*R*3	1.5	15	31	Porcine 2ZN	Adams *et al.* (1969), Baker *et al.* (1988)
4ZN	Rhombohedral	Hexamer	Dimer	*R*3	1.5	18	33	Porcine 4ZN	Bentley *et al.* (1976), Derewenda *et al.* (1988)
CUB	Cubic	Dimer	Monomer	$I2_13$	1.7	17	61	Porcine zinc-free	Dodson *et al.* (1978), Badger (1986)

A full list of the principal insulin crystal structures can be found in Derewenda *et al.* (1989), Table 1.

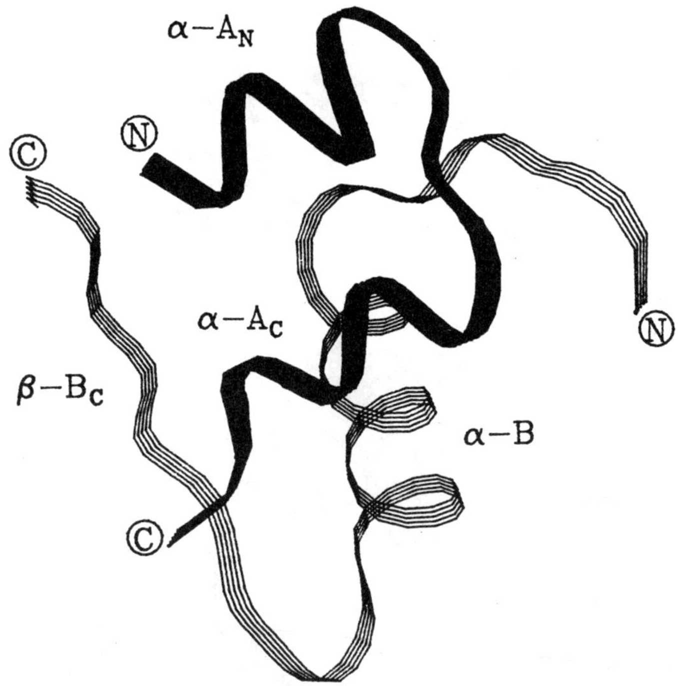

Figure 2.2 A schematic 'ribbon' representation of the C_α trace of the 2Zn insulin monomer. The A-chain (of 21 residues) is shown as a solid line. The B-chain (of 30 residues) is shown as a striped line. The *N*- and *C*-termini of the two chains are indicated by Ⓝ and Ⓒ respectively. The secondary structure elements are indicated:
α-A_N, A-chain *N*-terminal helix (residues A2–A8);
α-A_C, A-chain *C*-terminal helix (residues A12–A20);
α-B, B-chain helix (residues B9–B20);
β-B_C, B-chain *C*-terminal strand (residues B23–B29).
This general fold is representative of all native insulin conformers although considerable differences can occur at the *N*-terminus of the B-chain (may form an extension of the central B-helix) and to a lesser extent at the *N*-terminus of the A-chain and the *C*-terminus of the B-chain

variety of aggregation states) reveal the insulin molecule in a number of similar but distinct conformations; however, there is a general fold of the molecule (e.g. that in the 2Zn crystal form—see Table 2.1) that can be described (refer to Figure 2.2). The B-chain has a central α-helix (between B9 and B20), which at the *C*-terminal end turns into an extended strand. The A-chain has two regions of helix, at the *N*-terminus (A2–A8) and at the *C*-terminus (A12–A20), connected by a turn which makes the overall fold C-shaped. The two interchain disulphides occur at each end of the B-helix, with the A-chain straddling the B-helix and bringing together the sidechains from both chains which make up the buried non-polar core of the molecule: the *N*-terminal region of the B-chain either adopts an

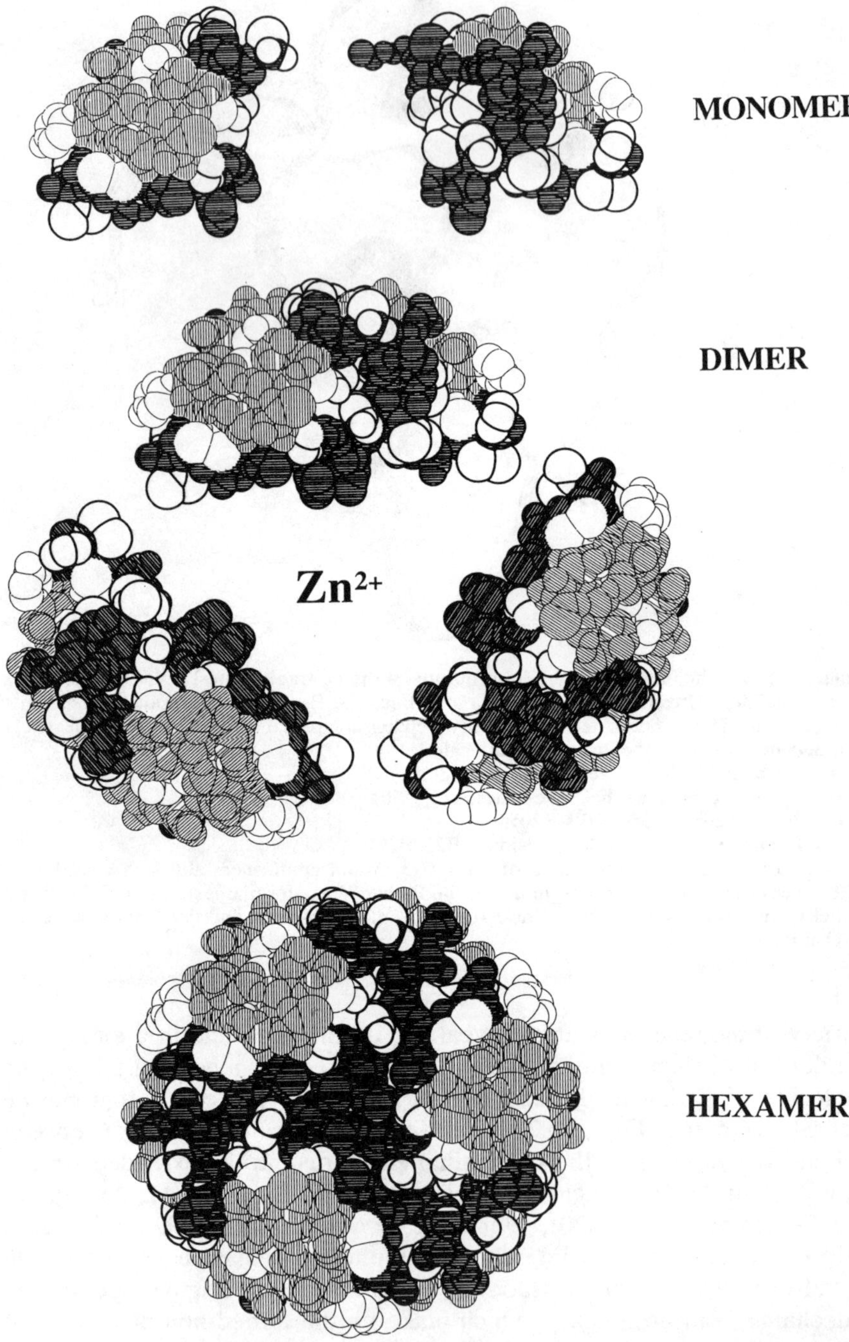
MONOMER
DIMER
Zn^{2+}
HEXAMER

extended conformation, or under certain conditions is an extension of the central B-helix.

The surfaces of the molecule will now be described (refer to Figure 2.3). The monomer has both polar and non-polar residues on its surface. The main polar surface is due to A-chain residues, but extends to the *N*-terminal residues of the B-chain, and along the exposed face of the B-helix. There is a large flat non-polar mainly aromatic surface comprising B-chain residues from the *C*-terminus and the B-helix which is the primary site of aggregation in dimers. There is another non-polar surface, with both aliphatic and aromatic residues from both chains, which is the site of dimer–dimer aggregation into hexamers. A full description of the insulin structure in the 2Zn pig insulin crystal is given by Baker *et al.* (1988). Accounts of the functional significance of aspects of insulin sequence and structure are discussed by Blundell *et al.* (1972) and Baker *et al.* (1988).

Atomic Coordinates

The atomic coordinates used as the basis of this study are those of 2Zn, 4Zn and cubic pig insulin (see Table 2.1). These structures contain five distinct conformers of insulin (two each for 2Zn and 4Zn, one for cubic) (see Figure 2.4). As each distinct monomer and dimer from each form was considered, eight simulations were performed (three dimer, five monomer). This results in a total of eleven distinct simulated monomeric conformers to be considered (six from the dimer simulations, five from the monomer). The codes for the various conformers are given in Table 2.2. All solvent molecules were excluded. In cases of multiple sidechain

Table 2.2 Codes for the conformers considered in this study

*Crystal form**	*Conformer†*	*Crystal conformer*	*Simulated conformer*	
			Isolated	*Associated*
2ZN	1	2ZX1	2ZM1	2ZD1
	2	2ZX2	2ZM2	2ZD2
4ZN	1	4ZX1	4ZM1	4ZD1
	2	4ZX2	4ZM2	4ZD2
CUB	1	CBX1	CBM1	CBD1/2‡

* As given in Table 2.1.
† Chinese convention (Peking Insulin Structure Group, 1971, 1974).
‡ For the cubic structure, there is only one crystal conformer; however, for the dimer simulation the initially identical monomers were arbitrarily named conformers 1 and 2.

Figure 2.3 Schematic view of the aggregation of insulin from the monomer to the dimer and further, in the presence of zinc ions, the assembly of dimers to form the hexamer. The atomic representation is space filling. Atoms of polar residues are shaded. The shading for the A-chain is thick lines, for the B-chain it is thin lines. The sidechains of non-polar residues are drawn as open spheres. The view is of the molecules of pig insulin in 2Zn crystals (see Table 2.1) down the crystallographic three-fold axis. Molecules in 4Zn insulin crystals also assemble to form hexamers in a similar arrangement, whereas in cubic crystals they form only dimers (for details of the crystal forms see Table 2.1)

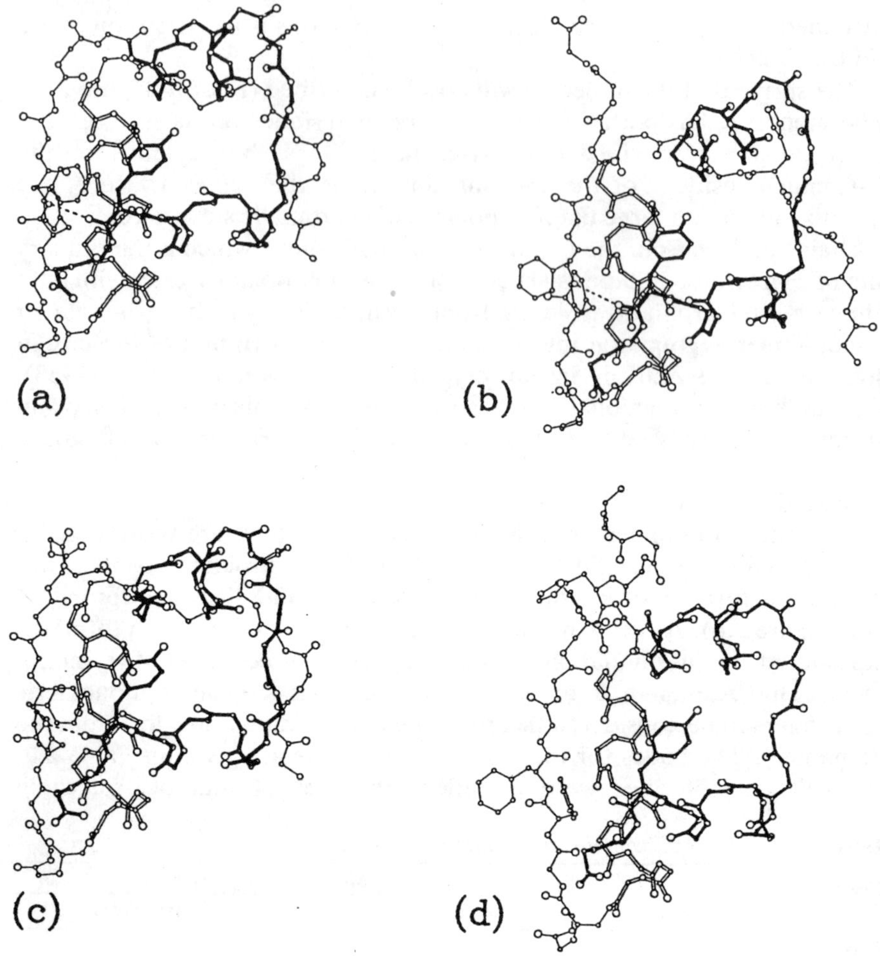

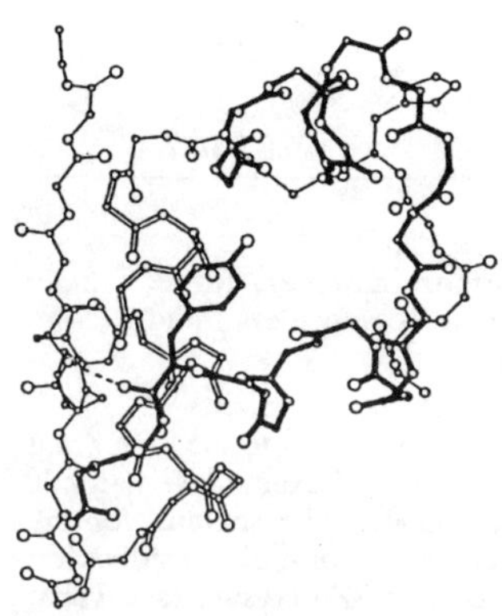

Figure 2.4 The five distinct pig insulin conformers considered in this study (see Table 2.1). The peptide backbone of the A-chain is drawn with thick bonds; that of the B-chain helix is drawn with open bonds. The remainder of the B-chain is drawn with thin bonds. The sidechains of B5 histidine, B24 and B25 phenylalanine and A19 tyrosine are also shown with the same bond type as their corresponding mainchain. The interchain hydrogen bond between the A19 carbonyl oxygen and the B5 backbone amide hydrogen is shown as a dashed line where it exists. (a) 2Zn molecule 1; (b) 2Zn molecule 2; (c) 4Zn molecule 1; (d) 4Zn molecule 2 and (e) cubic monomer. Note: in the view of the cubic structure B29 and B30 are missing; however, in this work, a more recent refinement model was used which has atomic positions for these regions (Badger, 1986)

conformations, the conformation adopted was either that with the highest occupancy or, if equally occupied, that which appears first in the coordinate set.

Simulation Methodology

Interaction Function and Molecular Representation

The simulations were performed using the CHARMM potential and molecular simulation package (Brooks *et al.*, 1983). The polar hydrogen representation was used as the molecular model (i.e. only hydrogen atoms that are potentially capable of participating in hydrogen bonding are included, the remainder being incorporated into extended heavy atoms). The CHARMM molecular topology and parameter sets used were TOPH19 and PARAM19 respectively. The hydrogen atom positions required to fulfil the representation were placed using the method of Brünger and Karplus (1988). In this representation the pig insulin monomer comprises 488 atoms.

For the electrostatic terms of the potential, a distance-dependent dielectric constant was employed (where ε_r is set to be numerically equal to the distance r, between two atoms in Å), as a correction for the shielding effect of the neglected solvent (Gelin, 1976; Weiner *et al.*, 1984). The non-bonded interactions were truncated at 9.0 Å, by means of a cubic switching function acting between 8.5 and 9.0 Å on an atom-pair basis (Brooks *et al.*, 1983). The pair list typically contained around 30 000 atom pairs. Ionizable amide and carboxyl groups were given full net electronic charges. Electrostatic interactions between 1 and 4 atom pairs were scaled by a factor of 0.5.

Molecular Dynamics Simulation Protocol

The models resulting from the refinement of the crystallographic data were subjected to energy minimization prior to molecular dynamics simulation. This was done in order to ensure that the structures were free of any major deviations from the reference parameters of the potential function. Such deviations in experimentally determined structures may include very poor bonded geometry (lengths, angles), inadvertent inversion of stereochemistry at chiral centres (L to D configuration) or atomic (van der Waals) overlap. These errors in the structure, if uncorrected, may result in a shock to the molecule in the initial stages of the simulation, and throw the resulting trajectory into an improbable region of conformational space.

It was found that the crystal coordinates were free of any gross errors and a short run of 10 steps of the steepest descents method was performed on the structures (both isolated and associated), which resulted in backbone (N, C_α, C) atom root-mean-square differences in atomic positions

(RMSDs) of less than 0.1 Å. An exception was the 4Zn structure which had poorer initial geometry than 2Zn or cubic. For this structure more extensive minimization runs of 100 steps of a conjugate gradients algorithm (Powell, 1977) were performed, which resulted in backbone RMSDs of less than 0.4 Å.

The MD simulations were performed at constant energy and volume, by integrating the Newtonian equations of motion for the systems on the potential surface described by the CHARMM interaction function and parameters. The Verlet (1967) algorithm was used with all bond lengths involving hydrogen atoms constrained (with a relative accuracy of 10^{-6}) with the SHAKE procedure (Ryckaert *et al.*, 1977; van Gunsteren and Berendsen, 1977), allowing an integration time step of 1×10^{-15} s. The non-bonded pair list was updated every 20 time steps.

The simulations were performed in three phases: heating, equilibration and production. In the heating phase, random velocities were assigned to the atoms from Gaussian distributions at successively increasing temperatures until reaching 300 K. In practice, this was done by initially assigning velocities at 50 K and increasing the temperature by 5 K every 200 steps over 10 ps. At each heating point, any net overall momentum of the system was removed. In the equilibration period the temperature of the system was monitored for 20 ps and if it exceeded a window of 10 K from 300 K the atomic velocities were reassigned at 300 K. In the production period, no temperature checking was performed and the integration proceeded uninterrupted up to 100 ps.

The simulations were performed using an optimized version of CHARMM (version 20) for a Cyber-205 supercomputer (B. Tidor, unpublished results) at the John von Neumann National Supercomputer Centre (JvNC) at Princeton University. Execution times were of the order of 1 CPU hour per 10 ps dynamics of the dimer.

GENERAL QUALITY OF THE SIMULATIONS

In order to investigate the general behaviour of the simulations certain overall properties of the molecules are monitored throughout the molecular dynamics trajectories. Measures of both molecular energy and structure are examined and are described below.

Energetics

The first 10 ps of the simulations is the heating stage, where the system is periodically injected with kinetic energy which correspondingly raises the total energy. This can be seen in the plot of the total and potential energy

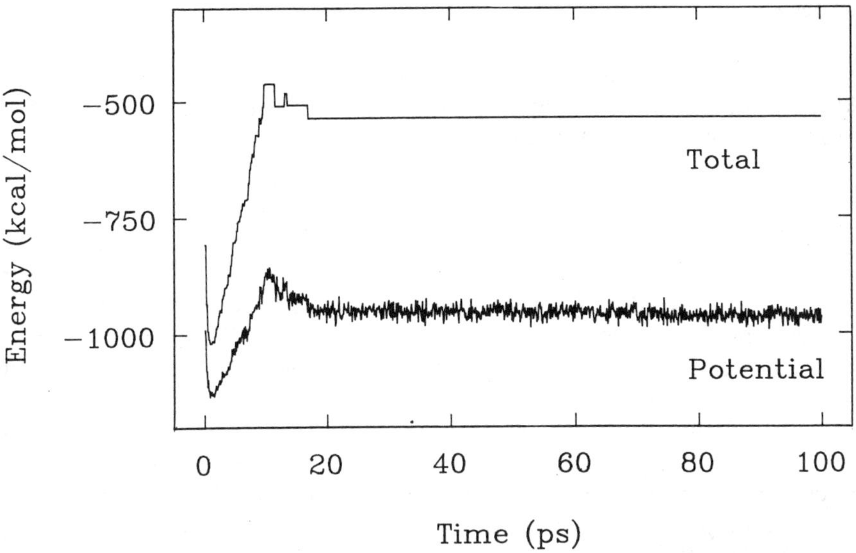

Figure 2.5 Time series of total and potential energy from the simulation of 2Zn molecule 1 in the isolated state. The initial heating phase is clearly visible by an increase in total energy. The conservation of total energy after equilibration is apparent, which is indicative of a stable trajectory. The potential energy stabilizes after about 20 ps. (1 kcal = 4.18 kJ)

time series for the isolated 2Zn molecule 1 simulation (Figure 2.5). After the 10 ps heating stage, the potential energy of the systems stabilizes after about another 10 ps. The 20 ps period required for the equilibration of energies observed in these trajectories is typical of protein dynamics simulations (Åqvist *et al.*, 1985).

The MD simulations were performed in the absence of external forces and therefore the total energy of the systems should be conserved. The ratio of the RMS fluctuations in the total energy to those in the kinetic energy are found to be around 1%, which is satisfactory for this type of calculation (van Gunsteren and Berendsen, 1977; Brooks *et al.*, 1983). A check was made for systematic drift in the total energy over time, by examining the gradient of the total energy time series (by least-squares fitting). From this analysis no significant gradients were found. The good conservation of energy is indicative of a stable integration, showing that the chosen time step was adequate and that the non-bonded interaction truncation was being effectively smoothed. The average temperature of the simulations is in the range 300–310 K, with RMS fluctuations of less than 10 K.

This analysis reveals that the energetics behave well by accepted criteria for this type of calculation and the simulations are therefore considered acceptable for further analysis.

Structure

Radius of Gyration and Molecular Volume

An overall measure commonly used to monitor changes to protein structure is the radius of gyration:

$$R_g^2 = \sum_{i=1}^{N} m_i(\boldsymbol{r}_i - \boldsymbol{r}_{cm})^2 \Big/ \sum_{i=1}^{N} m_i$$

where N is the number of atoms, m_i is the mass of atom i, $\boldsymbol{r}_i$ is the Cartesian position vector of atom i, and $\boldsymbol{r}_{cm}$ is the centre of mass of the molecule.

In Figure 2.6(a), the time series of R_g is shown for 2Zn molecule 1 in its simulation as an isolated species. The plot shows a decrease in R_g with time, indicating a contraction of the protein conformation, which amounts to ~4% of the R_g of the initial crystal conformer by the end of the simulation. The R_g of average structures generated from the simulations (over the 20–100 ps period) is on average 2.9–3% smaller than that of the corresponding crystal conformers (Table 2.3). The contractions found are comparable with those found in vacuum MD simulations of three different forms of lysozyme (average of 2.9–3%) (Post *et al.*, 1986) and the bovine pancreatic trypsin inhibitor (~5%) (Levitt and Sharon, 1988), proteins of similar size to insulin.

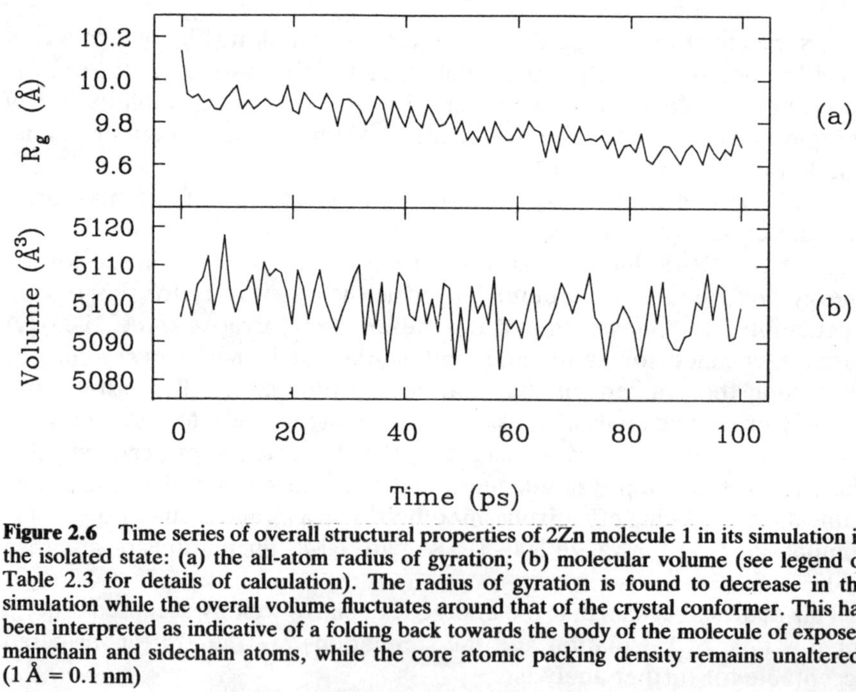

Figure 2.6 Time series of overall structural properties of 2Zn molecule 1 in its simulation in the isolated state: (a) the all-atom radius of gyration; (b) molecular volume (see legend of Table 2.3 for details of calculation). The radius of gyration is found to decrease in the simulation while the overall volume fluctuates around that of the crystal conformer. This has been interpreted as indicative of a folding back towards the body of the molecule of exposed mainchain and sidechain atoms, while the core atomic packing density remains unaltered. (1 Å = 0.1 nm)

Table 2.3 Overall properties of the crystal and simulated average conformers

Code*	R_g† (Å)	*Volume*‡ ($Å^3$)	$\langle \Delta r^2 \rangle^{1/2}$** (Å) A-chain	B-chain	*RMSD*†† (Å) All atom	Backbone
2ZX1	10.1	5096	0.85	0.76	—	—
2ZX2	10.3	5100	0.78	0.75	—	—
4ZX1	9.9	5077	0.85	0.80	—	—
4ZX2	10.7	5084	0.89	0.82	—	—
CBX1	9.9	5120	0.87	0.92	—	—
2ZM1	9.8	5106	0.48	0.51	1.98	1.13
2ZM2	10.0	5113	0.47	0.58	2.65	1.99
4ZM1	9.7	5107	0.46	0.54	2.32	1.62
4ZM2	10.4	5119	0.41	0.36	2.60	1.62
CBM1	9.6	5101	0.35	0.42	1.91	1.39
2ZD1	9.9	5103	0.45	0.66	1.29	0.92
2ZD2	9.9	5106	0.50	0.56	1.92	1.47
4ZD1	9.7	5107	0.61	0.64	1.98	1.60
4ZD2	10.3	5090	1.09	0.80	2.54	2.02
CBD1	9.7	5101	0.44	0.51	1.84	1.81
CBD2	9.6	5108	0.39	0.44	1.99	1.25

* As defined in Table 2.2; for simulated species, refers to average conformer.
† Radius of gyration calculated from all atomic positions.
‡ Volume calculated by examining atomic van der Waals' presence in a box of 40 $Å^3$ using volume elements of 0.28 $Å^3$. Atomic radii from CHARMM parameter set PARAM19 scaled by a factor of 0.9.
** RMS atomic displacements calculated from crystallographic *B*-factors or from simulated fluctuations.
†† RMS difference in atomic positions from initial crystal conformer, optimized on the backbone (N, C_α, C) atoms.
Note: average properties from simulations calculated over 20–100 ps period.

A decrease in R_g in simulations has been interpreted as indicative of an increase in packing density of the protein atoms via a simple spherical approximation to the shape of the protein, i.e. volume = $4\pi R_g^3$ (Weiner *et al.*, 1984). For insulin, which is not highly globular, this approximation is poor and the volume of the molecule was monitored directly through the simulations. The analysis revealed that the molecular volume shows no significant decreasing trend in the simulations. In the 2Zn molecule 1 simulation in the isolated state the range of molecular volumes sampled in a trajectory accounts for only 0.7% of the volume of its initial crystal conformer (Figure 2.6(b)). The volumes of the simulation-average conformers are within ~0.5% of their corresponding crystal conformers (see Table 2.3).

The analysis of the molecular radius of gyration and volume reveals that the conformation of the molecule contracts, but that this is not due to an increase in packing density in the protein interior. The contraction in R_g may be interpreted as a folding back of exposed regions of the peptide mainchain and/or exposed sidechains towards the body of the molecule

(Weiner *et al.*, 1984). This effect is due to the lack of interactions with the (neglected) surrounding solvent and crystal contacts and must be expected in protein MD simulations performed *in vacuo*.

Deviations from Crystal Structures

The deviations of the structures from their initial (crystal) conformations were monitored throughout the trajectory. This was performed by calculating the conventional RMS in atomic positions after superposition (RMSD) (Kabsch, 1976) and the RMS in atom–atom distances (R_d) for each transient structure vs. the initial crystal structure:

$$R_d = \left[\frac{1}{2N(N-1)} \sum_{i=1}^{N} \sum_{j<i} (r_{ij}^{A} - r_{ij}^{B})^2\right]^{1/2}$$

where N is the number of atoms and r_{ij}^{A} is the interatomic distance between atoms i and j in structure A.

In addition, the components of R_d, that constitute a difference-distance matrix (**D**), were examined to identify the specific structural origins of R_d:

$$\mathbf{D} = \left[r_{ij}^{A} - r_{ij}^{B} \right]$$

The distance matrix measures were chosen as they give an indication of the overall change in fold (Åqvist *et al.*, 1985) without relying on superposition reference frames (an advantage when comparing many conformations), and involve more terms than the RMSD ($\frac{1}{2}N(N-1)$ vs. N). The time series $R_d(t)$ were calculated for the C_α atoms of the monomers (both isolated and associated) in the simulations at 0.25 ps intervals throughout the trajectories.

In general, the deviations from the initial crystal structures stabilize after about 15–20 ps to RMS differences in C_α interatomic distances (R_d) of around 1.0–1.5 Å. An exception is the 4Zn molecule 2 associated species where the deviations increase up to around 40 ps and are more extensive (R_d of approximately 2.0 Å). A more detailed discussion of the results of this analysis is now presented by consideration of the trajectories of the 2Zn crystal form.

The plot of the time series of R_d for the 2Zn dimer simulation is shown in Figure 2.7(a). There is a clear difference in behaviour for the two constituent monomers when one examines the overall (both chains) deviations. At ~11 ps there is a steep transition in the time series for molecule 2 that develops a difference over molecule 1 of 0.5 Å (0.6 Å→ 1.1 Å) over a period of ~6 ps. The individual chain components of this value reveal that the change occurs predominantly in the B-chain. The C_α difference-distance matrix (**D**) of the 17 ps vs. 11 ps transient structures for molecule 2 (not shown) reveals that the last three residues of

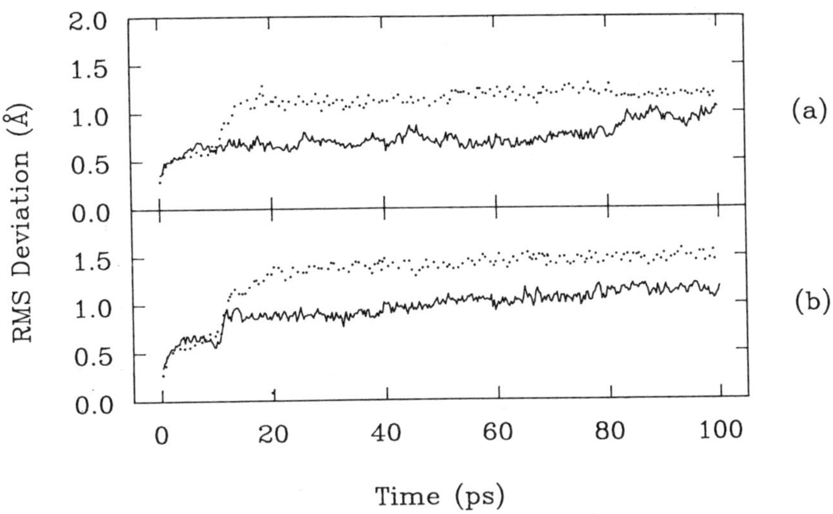

Figure 2.7 Time series of the deviation of 2Zn simulated conformers from their initial crystal conformation. The measure used for comparison is the RMS difference in inter-C_α distances (R_d) (1 Å = 0.1 nm): (a) associated species and (b) isolated species (molecule 1, solid line; molecule 2, dotted line). In the simulations molecule 2 conformers are generally found to deviate more from their crystal conformation than molecule 1 conformers

the B-chain (B28, B29, B30) move away from the A-chain *C*-terminal helix. Also evident is a closing up of the upper strand region of the B-chain (B25, B26, B27) against the B-helix. These changes are illustrated by the closing up of the *N*-terminus of the A-chain and the *C*-terminus of the B-chain seen in molecule 2 between 7 and 11 ps (Figure 2.8(a)).

In the simulations of the 2Zn conformers as isolated species, very similar behaviour is observed (see Figure 2.8(b)). A transition in the overall time series for molecule 2 at ~9 ps over a period of ~4 ps raises a 0.4 Å difference over the deviations for molecule 1, which is maintained for the remainder of the simulation. The transition in the isolated species has contributions from both A- and B-chains and again involves movements of the B-chain and the *N*-terminus of the A-chain, specifically a reduction of the separation of the A-chain *N*-terminus and the B-chain *C*-terminus (Figure 2.8(b)). These structural changes in the simulations are interesting as, in the crystal, the salt bridge between the *N*-terminus of the A-chain and *C*-terminus of the B-chain found in molecule 1 is not found in molecule 2 (Baker *et al.*, 1988). The origin of this difference in the conformation of the two molecules has been rationalized by examining the crystal packing of the hexamer (Chothia *et al.*, 1983), where the primary interhexamer contacts are found to involve largely the A-chain of molecule 2 (Baker *et al.*, 1988).

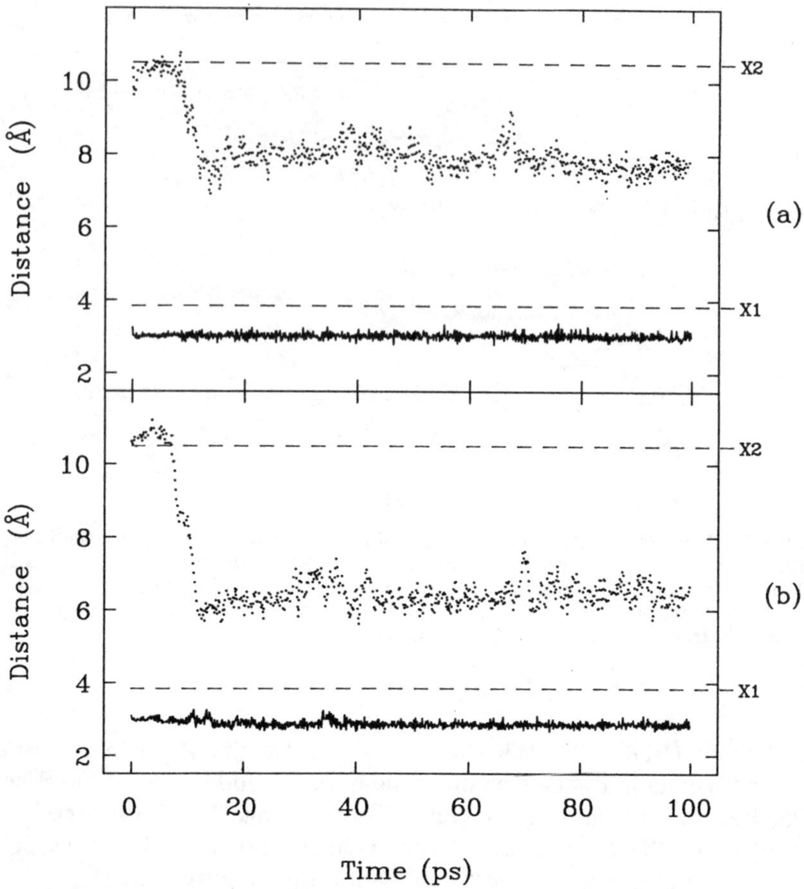

Figure 2.8 Time series of the distance between A1 peptide nitrogen and B30 carbonyl carbon in the 2Zn trajectories: (a) associated molecules and (b) isolated molecules (molecule 1, solid line; molecule 2, dotted line). The distances found in the 2Zn crystal conformers of molecule 1 (3.8 Å) and molecule 2 (10.5 Å) are indicated by dashed lines and are marked X1 and X2 respectively. (1 Å = 0.1 nm)

Large positional deviations in the A-chain are due to movements of A5 glutamine, a result of a change in the main chain dihedral angle ϕ (C–N–C_α–C) of A6 cystine (Figure 2.9(b)) between 13 and 17 ps, which moves from ~−65° to −115°. This is analogous to a change from the crystal conformation of molecule 2 (−66°) to molecule 1 (−101°) (Dodson *et al.*, 1980b). A similar change occurs in this conformer in the associated state, but the change is subsequently reversed (Figure 2.9(a)). The time series of R_d for the trajectory of the isolated 2Zn molecule 2 vs. the crystal conformer of 2Zn molecule 1 reveals only a small decrease in R_d with time (not shown), indicating that no significant overall convergence of the

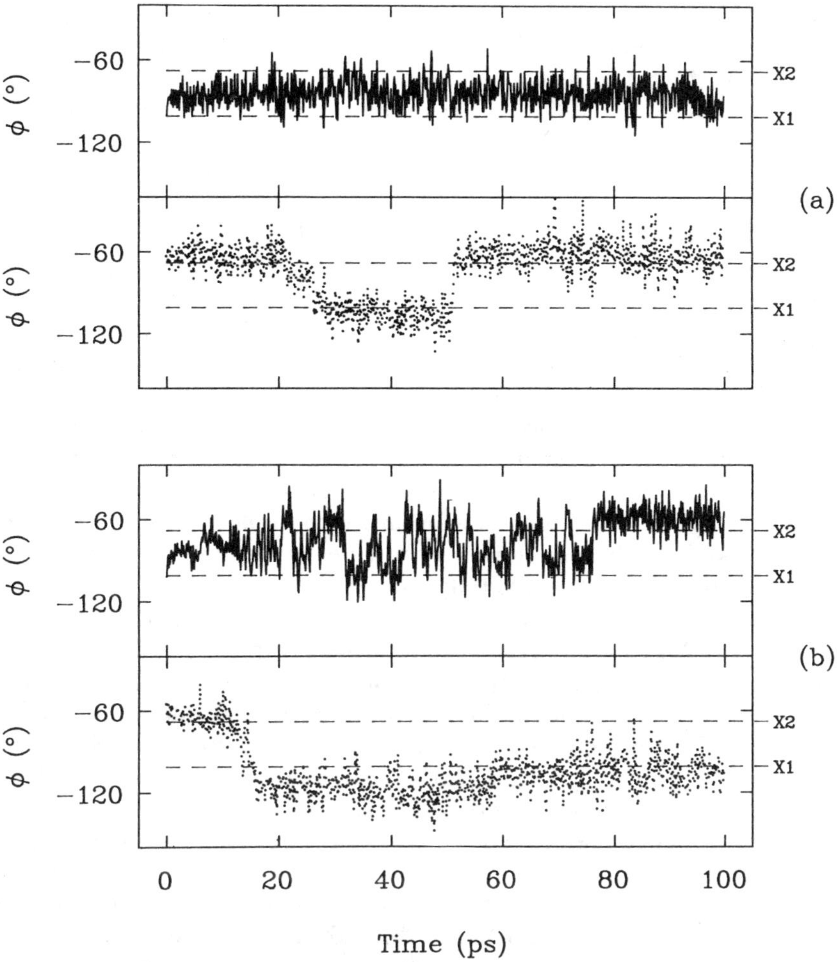

Figure 2.9 Time series of the peptide backbone dihedral angle ϕ (C–N–C_α–C) in A6 cystine in the 2Zn trajectories: (a) associated molecules and (b) isolated molecules (molecule 1, solid line; molecule 2, dotted line). The angles found in the 2Zn crystal conformers of molecule 1 (−101°) and molecule 2 (−66°) are indicated by dashed lines and are marked X1 and X2 respectively

conformational differences is occurring, behaviour which has been observed in other MD simulations of 2Zn insulin (Krüger *et al.*, 1987).

In the simulations, where the conformers are freed from the packing constraints of the crystal, the separation between chain termini in molecule 2 is found to decrease and move towards that found in molecule 1. This change is not surprising owing to the attractive electrostatic interactions between the opposite net charges of the termini. This event has also been observed in MD simulations of isolated 2Zn molecules 1 and 2 using a different interaction potential (Krüger *et al.*, 1987). In the isolated state, molecule 2 is found to change its A-chain *N*-terminal conformation towards that found in the crystal for molecule 1. This has been observed in energy minimization studies of the 2Zn insulin conformers (Wodak *et al.*, 1984).

Summary of Overall Structural Consequences of the Simulations

Structures undergo large (up to 1 Å in R_d) and sometimes rapid (less than 1 ps duration) structural transitions that have been traced to conformational features which may be considered to be the most sensitive to the change of environment from the crystal. The deviations from the initial structure are comparable with previous *in vacuo* MD simulations of proteins (Åqvist *et al.*, 1985; Levitt and Sharon, 1988, and references therein). The deviations are low especially when it is considered that the aggregation states (hexamer for 2Zn and 4Zn, dimer for cubic) found in the crystal are not being modelled explicitly in the simulations. An exception is the cubic dimer simulation which models the crystal association state and which does show very low deviations with an overall R_d of ~1.0 Å, and in the A-chain of only ~0.5 Å.

In most simulations the deviations from the crystal structures stabilize after about 20 ps; thus average conformers for the simulations were generated over the 20–100 ps (see section on Interconformer relationships). The conventional RMS differences in equivalent atomic positions (RMSDs) for these average structures vs. their corresponding crystal conformer are given in Table 2.3. The rank (increasing) order of the deviations for the crystal forms within association classes is 2Zn, cubic, 4Zn. Conformers simulated in the isolated state generally deviate more than those in the dimer.

The low deviations of the trajectories from the crystal conformations may be viewed as indicative of the fact that the overall insulin fold is stable, even with the omission of solvent and counterions from the model. There is evidence that this is not the case for the 4Zn molecule 2 conformer, as large deviations are found to occur in its simulation in the dimer association state (see Figure 2.17(c)). This perhaps is not surprising when one considers that the particular characteristics of the 4Zn molecule 2 conformer (extension of the B-chain helix at the *N*-terminus and concomitant increase in A- and B-chain separation—see Figure 2.4) have been shown to be sensitive to surrounding medium effects (Bentley *et al.*, 1978).

In conclusion, the simulations have produced stable trajectories in which the overall structures of the molecules have not deviated greatly from those experimentally observed. The specific nature of the structural changes observed in the simulations relates to the major differences between the crystal conformers. Many of the major structural changes are likely to be consequences of the neglect of the crystal environment in the simulations.

The trajectories can therefore be drawn upon as giving representative samples of the conformational fluctuations available to the insulin molecule, at least on the picosecond timescale, and in the next sections are analysed at the atomic and secondary structure level.

ATOMIC MOTION

The atomic motions in the insulin conformers are examined on the basis of experimentally determined crystallographic thermal parameters (*B*-factors) and from atomic fluctuations calculated from the simulated trajectories. Both the magnitude and spatial distribution of the atomic fluctuations are considered. The study extends to an examination of the collective character of the atomic motion in the simulations in order to compare with results from an interpretation of X-ray diffuse scattering experiments on insulin crystals (Caspar *et al.*, 1988).

Amplitudes of Motion

The magnitude and distribution of atomic fluctuations in proteins may be determined from X-ray diffraction data (Ringe and Petsko, 1985). In the Debye–Waller treatment, the thermal vibrations of atoms are assumed to be harmonic (Willis and Pryor, 1975). The thermal parameter (*B*) is related to the mean square fluctuations of an atom by

$$B = \frac{8}{3}\pi^2\langle \Delta r^2 \rangle$$

where Δr^2 is the mean-square radial displacement $(\Delta x^2 + \Delta y^2 + \Delta z^2)$, and $\langle\rangle$ denotes a time and space average.

In this treatment displacements are assumed to be isotropic $(\Delta x^2 = \Delta y^2 = \Delta z^2)$. Anisotropic thermal parameter refinement, which results in a mean square displacement tensor (**B**-matrix) for each atom, is not at present available for insulin. The observed mean square displacements may be regarded (although the decomposition is somewhat arbitrary) as having components from intramolecular motion $(\langle \Delta r^2 \rangle_{int})$, rigid-body lattice motions $(\langle \Delta r^2 \rangle_{lat})$ and static disorder $(\langle \Delta r^2 \rangle_{dis})$ in the crystal:

$$\langle \Delta r^2 \rangle = \langle \Delta r^2 \rangle_{int} + \langle \Delta r^2 \rangle_{lat} + \langle \Delta r^2 \rangle_{dis}$$

The *B*-factor therefore represents a measure of the protein dynamics, and has been found to correlate well with notions of secondary, tertiary and quaternary structure (Ringe and Petsko, 1985; Artymiuk *et al.*, 1979).

In general, the magnitude of the backbone (N, C_α, C) atomic fluctuations in the simulations are lower (by ~30–50%) than those derived from crystallographic *B*-factors (see Table 2.3). Very similar results were obtained in MD simulations of 2Zn insulin conformers using a different potential function (Krüger *et al.*, 1987). Simulation of conformers in a dimer environment generally results in larger amplitudes of motion than when isolated, perhaps reflecting some cooperative effects. The fact that the atomic fluctuations in the simulations are smaller than found in the crystal is understandable as the simulations relate only to internal protein motion. As mentioned above, the crystal *B*-factors include the effects of lattice motions and/or static disorder which are not accounted for in the simulations (Levitt and Sharon, 1988). Thus, if atomic fluctuations from the simulations are considered representative of the true internal protein motions (on the picosecond timescale), the proportion of the *B*-factor derived fluctuations not accounted for by the simulations may then be regarded as indicative of the lattice motion/static disorder contribution to the observed atomic motion.

From the simulations, the *B*-chain fluctuations are found to be larger than those in the A-chain, the opposite of that found in the crystal. This may reflect inadequacies in the simulations. In the crystal forms considered here, the A-chain is largely surface exposed. As previously mentioned, in the rhombohedral crystal forms (2Zn and 4Zn) the A-chain is involved in interhexamer contacts (Baker *et al.*, 1988). Crystal contacts are neglected in the simulations and this could, together with the omission of solvent, account for the discrepancy in the relative magnitude of the A- and B-chain fluctuations.

It is interesting to note that solution NMR studies of monomeric insulin reveal that the B-chain *C*-terminal β-sheet is very flexible (Weiss *et al.*, unpublished results). Unfortunately the NMR data are very preliminary and do not allow for a detailed comparison with simulation results at present.

Spatial Distribution

The spatial distributions of the atomic fluctuations in the simulations are compared with those derived from the *B*-factors resulting from the crystallographic refinements (Konnert and Hendrickson, 1980). This was done by examining the trends of the residue-averaged backbone (N, C_α, C) atomic fluctuations. In general, the residue-by-residue trends from the simulations are in poor agreement with those found in the crystal (Table 2.4). It was expected that, as the dimer is the lowest common association state of the crystal forms, the distribution of atomic motion of conformers simulated in a dimer environment would be in better agreement with the trends in crystal conformers than the corresponding simulation of the

Table 2.4 Correlation of the trends of atomic fluctuations in the simulations with those of the initial crystal conformers

	Correlation coefficient (R)			
	A-chain (21 residues)		*B-chain (30 residues)*	
*Crystal conformer**	*Simulation state Isolated*	*Associated*	*Simulation state Isolated*	*Associated*
2ZX1	0.14	0.51	0.67	−0.01
2ZX2	−0.44	0.24	0.35	0.19
4ZX1	0.05	0.30	0.33	0.56
4ZX2	0.47	−0.59	0.60	−0.47
CBX1	0.80	−0.03	0.67	0.72
CBX2	—	−0.13	—	−0.22

* For code see Table 2.2.
Residue-averaged backbone (N, C_α, C) results.
The correlation coefficient (R_{ij}) for the two structures i and j is

$$R_{ij} = \frac{\langle (B_i - \langle B_i \rangle)(B_j - \langle B_j \rangle) \rangle}{\langle (B_i - \langle B_i \rangle)^2 \rangle^{1/2} \langle (B_j - \langle B_j \rangle)^2 \rangle^{1/2}}$$

where: B_i are the residue-averaged fluctuations (B-factors) in structure i and $\langle B \rangle$ denotes mean crystallographic B-factor for the backbone atoms.

Table 2.5 Correlation matrix of the trends of atomic fluctuations in the conformers of 2Zn, 4Zn and cubic insulin derived from crystallographic B-factors

Code	*2ZX1*	*2ZX2*	*4ZX1*	*4ZX2*
A-chain (21 residues)				
2ZX2	0.41			
4ZX1	0.39	−0.08		
4ZX2	0.57	0.69	0.23	
CBX1	0.72	0.79	0.30	0.67
B-chain (30 residues)				
2ZX2	0.73			
4ZX1	0.73	0.76		
4ZX2	0.74	0.86	0.74	
CBX1	0.54	0.90	0.78	0.79

Residue-averaged (N, C_α, C) results.
The data are from the structures in Table 2.1 and the conformers are coded as given in Table 2.2.
The correlation coefficient is computed as given in the legend to Table 2.4.

isolated conformer. This is not generally found; for example, the isolated simulation of the cubic monomer reproduces the trends of motions in the dimeric crystal better than the dimer simulation (Table 2.4 and Figure 2.10(a)).

In general, the trends of atomic fluctuations in the crystal are in better

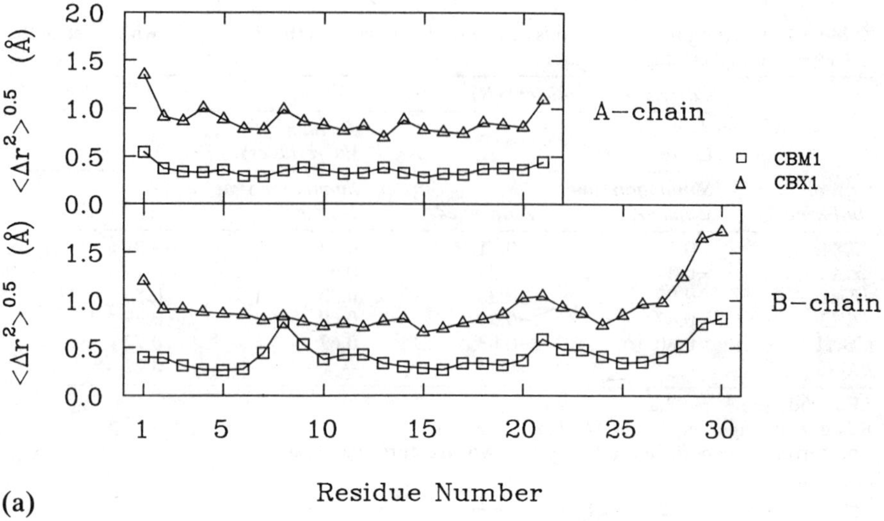

(a)

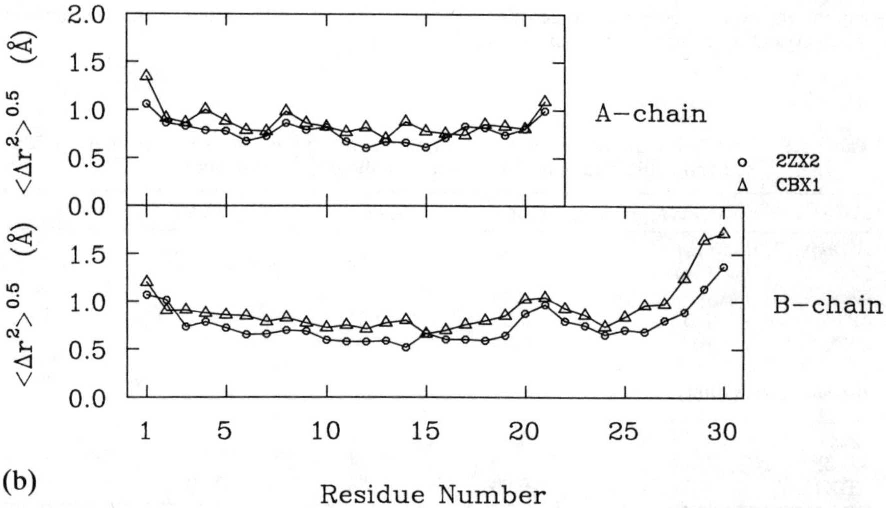

(b)

Figure 2.10 Comparison of residue-averaged backbone (N, C_α, C) atomic fluctuations: (a) simulation-average cubic conformer in the isolated state (CBM1) vs. its initial crystal conformer (CBX1); (b) crystal conformers of cubic (CBX1) and 2Zn molecule 2 (2ZX2). (Crystal values derived from isotropic *B*-factors; simulation results from 20–100 ps period.) (1 Å = 0.1 nm)

agreement with the simulation results in the B-chain than in the A-chain (see Table 2.4). Further, an examination of the residue-averaged trends in all of the simulations reveals that those of the B-chain are more similar between species than those of the A-chain (not shown). Interestingly, the similarity of the trends between the different crystal conformers is higher for the B-chain than the A-chain (Table 2.5). However, since the variation of the crystal trends is small, the correlation coefficients must be interpreted cautiously. Surprising similarities can be found. For example, in the B-chain the cubic conformer and 2Zn molecule 2 are most similar (correlation coefficient, $R = 0.90$) (see Figure 2.10(b)) whereas the cubic and 2Zn conformers of molecule 1 are least similar ($R = 0.54$). The greater variation of the trends of the A-chain motions between species compared with the B-chain may be due to the fact that in the dimer it is the B-chain that is principally involved in the monomer–monomer interface, which means that it is effectively in a more conserved environment between crystal forms than the A-chain. Furthermore, it has been noted that crystal packing contacts result in distinct differences in atomic mobility in the A-chain between molecules 1 and 2 in the 2Zn insulin crystal (Baker *et al.*, 1988).

Collective Nature

The collective behaviour of atomic displacements in the trajectories was investigated by examining the time correlation of the displacement of the residue or backbone atom centroids (McCammon, 1984):

$$\mathbf{R} = \left[\frac{\langle \Delta r_i \, \Delta r_j \rangle}{\langle \Delta r_i^2 \rangle^{1/2} \, \langle \Delta r_j^2 \rangle^{1/2}} \right]$$

where Δr_i is the instantaneous displacement of the centroid of the atoms of residue i from its mean position and $\langle\rangle$ denotes the time average.

The analysis yields a residue–residue correlation matrix **R**, which may be interpreted as follows. A high correlation coefficient (close to one) implies that two residues are displaying collective character in their motions. A low correlation coefficient (close to zero) implies that there is little coupling in the motions of the two residues.

The backbone (N, C_α, C) centroid displacement cross-correlations in the 2Zn dimer trajectory are shown in Figure 2.11(a) as a two-dimensional contour plot. The range of absolute (off-diagonal) correlation coefficients is 0.72–0.99, indicating that the motion of the whole molecule is highly coupled. Higher correlations occur along the diagonal, reflecting coupling between adjacent residues. More extensive regions with high correlations are found to occur, for example in region 29–39 (B8–B18)—the molecule 1 B-helix. This reflects the collective character of the motion of this helix. High correlations are also found to occur in off-diagonal positions, reflecting coupling of non-adjacent or non-contiguous residues. For example, between the two molecules of the dimer, a coupled motion is observed between the B-helix molecule 1 and the loop region 90–94 (B17–B21) in molecule 2.

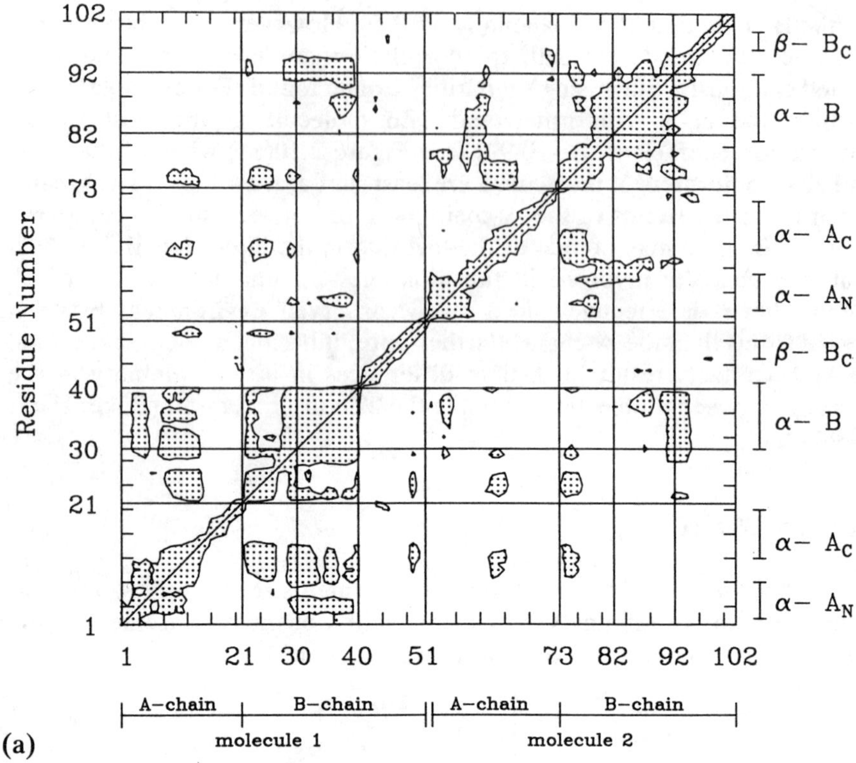
Residue Number
102
92
82
73
51
40
30
21
1
1 21 30 40 51 73 82 92 102
β– B$_C$
α– B
α– A$_C$
α– A$_N$
β– B$_C$
α– B
α– A$_C$
α– A$_N$
A–chain
B–chain
molecule 1
A–chain
B–chain
molecule 2

(a)

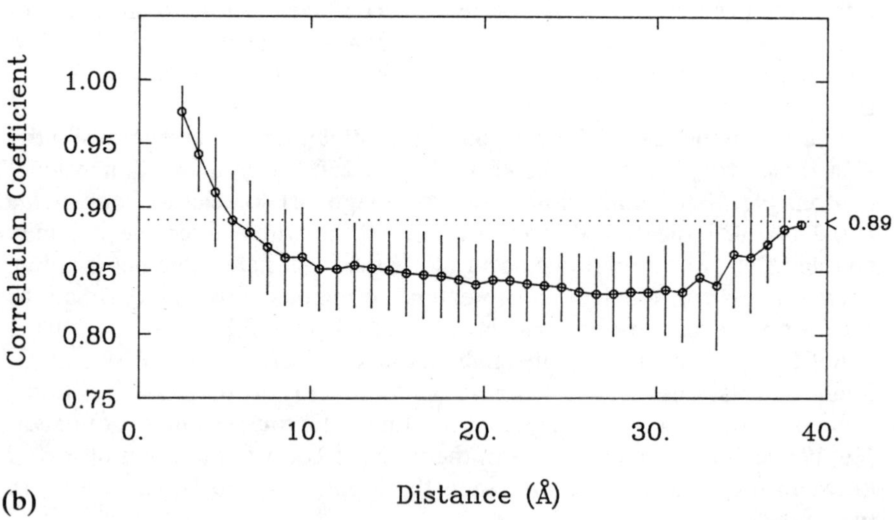
Correlation Coefficient
1.00
0.95
0.90
0.85
0.80
0.75
0. 10. 20. 30. 40.
< 0.89
Distance (Å)

(b)

These results may be compared with the findings of an interpretation of the thermal diffuse X-ray scattering (TDS) from 2Zn insulin crystals (Caspar *et al.*, 1988). TDS can give information on coupled correlated motion in crystals (e.g. Doucet and Benoit, 1987). In the study by Caspar *et al.*, the diffuse scattering was considered to arise from two main components, a variational term from internal protein motions and a lattice term from the rigid-body displacements of neighbouring molecules in the crystal lattice. The diffuse scattering was modelled by the use of two parameters, δ, the average magnitude of fluctuation, and γ, the mean coupling distance of atoms, which can be independently estimated from experimental measurements. Caspar *et al.* (1988) found that the variational scattering was best described by atomic fluctuations of the order of 0.69–0.78 Å coupled over an average distance of 4–8 Å. The lattice term was found to have fluctuations of ~0.25 Å over a distance of 20–30 Å. Thus it was found that the intramolecular atomic displacements account for 60–80% of the overall observed displacements.

Data from the simulation of the 2Zn dimer may be compared with this experimental study. Overall (average of both molecules and both chains), it is found that the magnitude of the atomic fluctuations in the simulations accounted for ~70% of those found in the crystal. From the analysis of the temporal correlation of the (backbone and residue) centroid fluctuations, the average distance over which residues are coupled is estimated to be 6–8 Å (see Figure 2.11(b)). Although the interpretation of TDS data from protein crystals is still in its infancy, the agreement of simulation with experiment is encouraging.

In the TDS study described, the analysis was based upon one still X-ray photograph of insulin and was therefore unable to resolve the three-dimensional distribution of the coupled displacements; however, this is

---

Figure 2.11 Collective atomic motion in the 2Zn dimer simulation over the period 20–100 ps monitored by the cross-correlation of backbone (N, C_α, C) atom–centroid displacements. (a) Residue–residue correlation matrix represented as a two-dimensional contour plot. The range of absolute (off-diagonal) correlation coefficients is 0.7–0.99. A single contour is drawn at a value of 0.89 which corresponds to one standard deviation above the mean of 0.85. Larger correlations (the dotted areas within the contour lines) occur along the diagonal and within the closed-off regions off the diagonal (regions close in the amino acid sequence); however, significant correlations also occur in off-diagonal positions. The residue ranges of the secondary structure elements found in the crystal are indicated to the right of the plot (refer to Figure 2.2). (b) The distance dependence of the collective motion of the backbone atoms is shown by plotting correlation coefficient vs. the average distance between backbone centroids. The correlation coefficients are averaged over 1 Å ranges, and are plotted at the centre of the range, with uncertainties of one standard deviation of the sample distribution for that range. A line is drawn at a value of 0.89 which corresponds to one standard deviation above the mean of 0.85. The highly coupled motion occurs within a range of less than 5–6 Å. Data for the residue centroids follow a very similar trend but are shifted (by about 1–1.5 Å) to longer distances. (1 Å = 0.1 nm)

available from the simulations. The simulations reveal different types of collective behaviour. Short-range correlation of adjacent residues occurs as a result of a coupling through the covalent backbone linkage. Longer-range coupling is found to occur in regions of defined secondary structure, for example α-helices, which suggests that the spatial restraints on backbone atoms imposed by the helical hydrogen bonding give rise to their collective motion. A further class of collective motion occurs where the packing of secondary structure brings sidechains into close proximity (for example at the interface between monomers). Motions of these secondary structure regions force concomitant movements of the atoms at the packing interface to accommodate the motion (Chothia *et al.*, 1983). It is the motion of secondary structure regions that is investigated in the next section.

SECONDARY STRUCTURE MOTION

In this section the secondary structure motion in the insulin molecule in the simulated trajectories is examined. Overall structural behaviour of the secondary structure is characterized and is followed by an examination of the relative motions of certain secondary structure regions. The relationship of the secondary structure dynamics in the simulations and in the crystal is addressed. First, some existing studies of the motion of secondary structure in proteins are mentioned.

Studies of Secondary Structure Motion

The motion of secondary structure elements (in particular α-helices) in proteins has been examined both in interpretive studies of protein crystal structures and in dynamic simulations. An examination of the crystal structures of the globin series (Lesk and Chothia, 1980) revealed that equivalent adjacent helices in the homologous (20–30% sequence) structures had relative displacements ranging up to 2.5 Å and orientations up to ~10°. Identical (backbone atom RMSDs of less than 0.2 Å) helices found in non-equivalent crystal environments, as in the case of 2Zn and 4Zn insulin, show relative displacements of <1.5 Å and orientations of typically 5–15°, but as large as 39° (Chothia *et al.*, 1983).

From the earliest protein dynamics simulations, collective behaviour of certain regions of atoms has been found, by examination of both the correlation functions of atomic displacements (Swaminathan *et al.*, 1982) and residue-centroid displacements (McCammon, 1984). Later studies have shown collective behaviour particularly associated with helical regions of secondary structure. In an MD simulation of the *C*-terminal fragment of the L7/L12 ribosomal protein (Åqvist *et al.*, 1985), correlated motions were found both within and between the secondary structure elements. In an MD study of lysozyme the relationship between atomic displacements and secondary structure was examined and revealed two main types of behaviour (Post *et al.*, 1989). For helices, atomic displacements were found to involve the entire secondary structure element, whereas for loop regions the displacements were more localized, involving mainly correlated changes in dihedral angles. In a study of a long (300 ps) trajectory of myoglobin the helix packing was examined in minimized structures sampled from the trajectory (Elber and Karplus, 1987). Adjacent helices were found to have relative displacements of 0.4–0.8 Å, and orientations in the range 1–5°.

In these studies the collective motion of helices which retain a relatively invariant internal structure has led to a view of them as being rigid or semi-rigid in nature. However, studies of helix geometry and hydrogen bonding patterns in known protein structures have revealed that helices are more deformable than previously thought (Barlow and Thornton, 1988; Baker and Hubbard, 1984).

Characterization of Secondary Structure Dynamics

In this study, the method of examining the intrinsic structural integrity of protein secondary structure uses a distance matrix approach (Havel *et al.*, 1983; Elber and Karplus, 1987). The distances between particles i and j, r_{ij}, in a given conformation make up a matrix **R**. Of interest is the time-averaged fluctuation of these distances in the simulation (Caves, 1989b):

$$\mathbf{DD} = \langle (\mathbf{R}_t - \langle \mathbf{R} \rangle)^2 \rangle^{1/2}$$

where $\langle \rangle$ denotes the time average and the subscript t denotes the value at time t.

This analysis yields a residue–residue variance (actually RMS) matrix **DD**, which may be interpreted as follows. A low element (close to zero) means there is little variation in the specific interatomic (e.g. C_α) distance, and thus the structural element is relatively rigid. A high value means that there is a large variation in the interatomic distance and that the structural element is flexible in nature.

The **DD** matrix (the RMS difference-distance matrix) for the C_α atoms was calculated from each of the simulations over the period 20–100 ps at a time resolution of 0.1 ps. In Figure 2.12 the **DD** matrix for the isolated 2Zn molecule 1 simulation is shown as a two-dimensional contour plot. Along the leading diagonal are found the self–self (zero) terms. As expected, low fluctuations (less than 0.2 Å) are found for residues adjacent in the

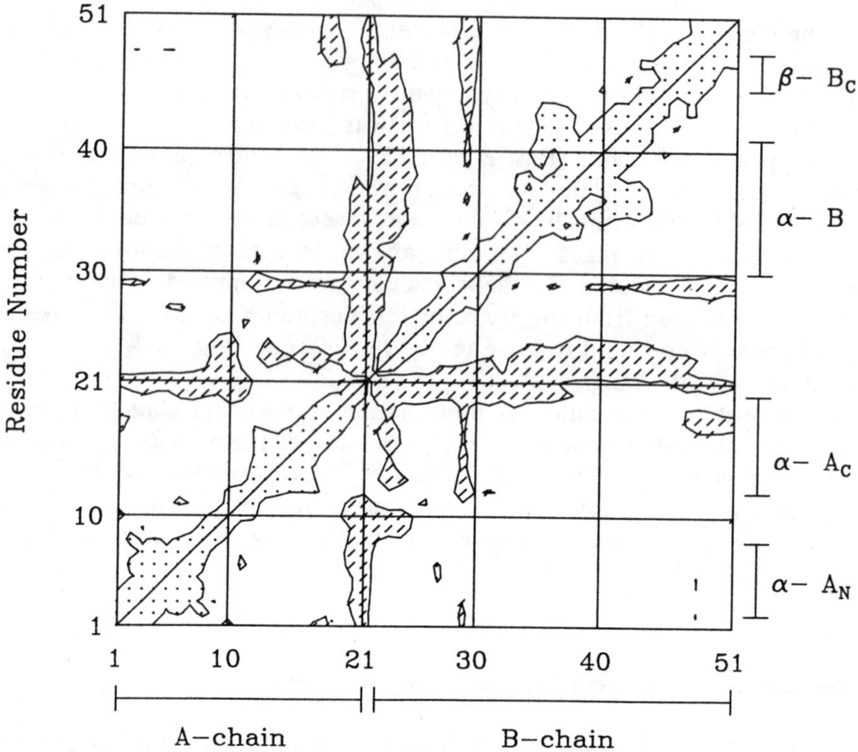

Figure 2.12 The RMS variation in inter-C_α distances in the 2Zn molecule 1 isolated trajectory over the period 20–100 ps, represented as a two-dimensional contour plot. Contours are drawn at values of 0.18 Å (lower values indicated by dots) and at 0.58 Å (higher values indicated by diagonal dashing) which correspond to one standard deviation below and above the mean of (off-diagonal) elements of 0.38 Å (1 Å = 0.1 nm).

Low variation occurs mainly along the diagonal (residues near in sequence) and is more extensive in helical regions. High variation is found to occur at loop regions and chain termini. The ranges of secondary structure elements found in the crystal are indicated on the right of the plot (refer to Figure 2.2)

sequence (commonly extending up to 2 residues either side), reflecting the constraints of covalent linkage. These regions manifest themselves as low contours about the leading diagonal. There are also extensive contiguous regions of secondary structure that may be regarded as semi-rigid in nature. These regions include the observed secondary structure elements and reflect the spatial restraints of the interactions (e.g. hydrogen bonds) that give rise to them. Non-contiguous regions may also be constrained. These may be explained by covalent (S–S bridge) or non-covalent (hydrogen bond, salt bridge) interactions. Flexible regions (such as the *N*-terminus of the B-chain) are revealed clearly, and tend to display flexibility with respect to the whole of the rest of the molecule.

Relative Motion of Secondary Structure Regions

Certain regions were selected to serve as probes of the internal dynamics of the molecule via the monitoring of their relative motion. The regions selected were AN, AC, BH and BS (as defined in Figure 2.13) which coincide approximately with the *N*-terminal A-helix, *C*-terminal A-helix, B-helix and *C*-terminal B-chain strand respectively. These regions are amongst those found to have high intrinsic structural integrity throughout the simulations, with overall RMS fluctuations of their internal C_α distances of less than 0.2 Å. The time series of the relative displacements of the centroids of the four selected structural regions show RMS fluctuations of 0.2–0.3 Å and ranges of 1.0–1.5 Å. The relative orientation of the secondary structure regions was taken to be the inner angle between their axes. Instantaneous axes were defined by calculating the principal axes of inertia of the regions by diagonalization of the second moment matrix of the centre of mass Cartesian position vectors of their backbone (N, C_α, C) atoms (Caves, 1989b). The orientational motions typically have RMS fluctuations of 2–3° with ranges of 15–20°. The BH,AC and BH,BS displacement and orientational motion for the cubic and 2Zn molecule 1 isolated simulations is shown in Figure 2.14(a). It was found that there is a high degree of overlap of the secondary structure motions between the trajectories. Thus the molecular dynamics are sampling very similar

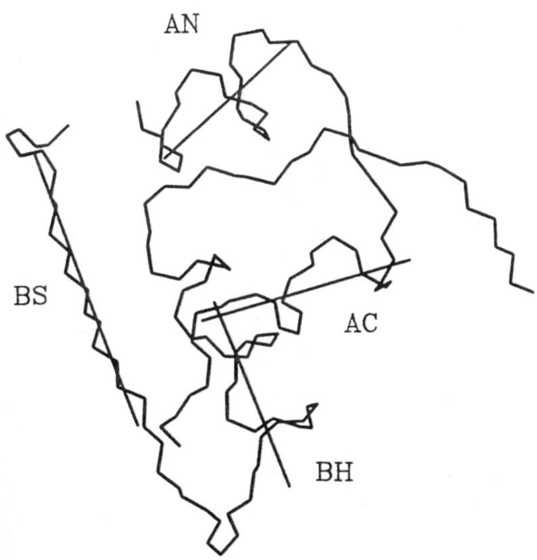

Figure 2.13 Axes defined for selected secondary structure regions, illustrated for the 2Zn molecule 1 crystal conformer. The selected regions are as follows: AN, region A2–A9; AC, region A13–A19; BH, region B12–B20; BS, region B24–B28. These regions were all found to have high intrinsic structural integrity in the trajectories

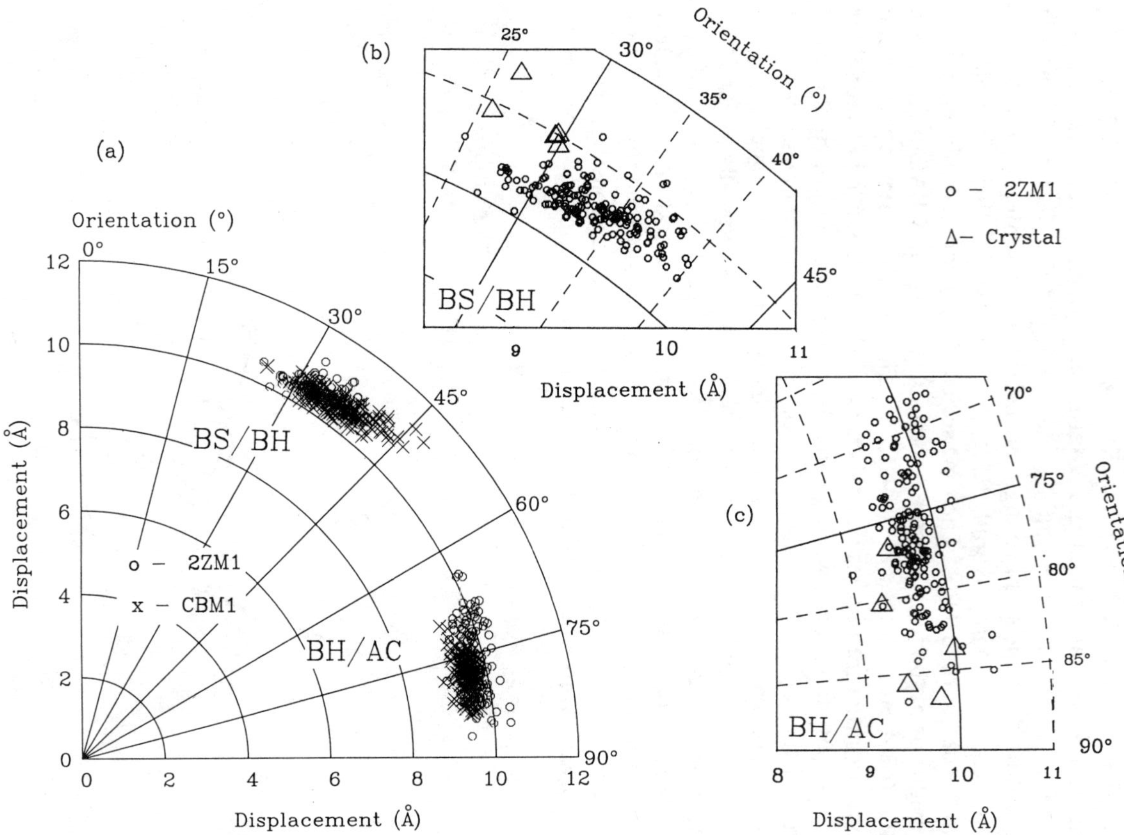

Figure 2.14 Relative motions of selected secondary regions. The instantaneous orientation vs. displacement is plotted for the 20–100 ps regions of the trajectories at 0.5 ps intervals. The codes for the regions can be found in the legend of Figure 2.13. For details of calculation see text. (a) BS/BH and BH/AC motions in the cubic isolated (○) and 2Zn molecule 1 isolated (×) trajectories. (b) Close-up of the BS/BH region for the 2Zn molecule 1 isolated trajectory. The orientation and displacement of these regions in the five crystal conformers (Figure 2.2) are shown (Δ). (c) Close-up of the BH/AC region

Table 2.6 Ranges of the relative packing of secondary structure regions in the crystal and in one simulated trajectory

	Crystal			*Simulation*†		
*Elements**	*Minimum*	*Maximum*	*Range*	*Minimum*	*Maximum*	*Range*
Displacement (Å)						
AC,AN	11.69 (CBX1)	12.45 (2ZX2)	0.76	10.95	12.61	1.66
BH,AN	15.23 (4ZX1)	15.35 (2ZX2)	0.12	14.37	15.88	1.51
BH,AC	9.27 (2ZX1)	9.99 (CBX1)	0.72	9.02	10.54	1.52
BS,AN	12.14 (4ZX1)	13.63 (4ZX2)	1.49	11.82	13.03	1.21
BS,AC	11.20 (CBX1)	13.26 (4ZX2)	2.06	10.18	11.70	1.51
BS,BH	0.90 (CBX1)	11.41 (4ZX2)	0.51	10.04	11.33	1.30
Orientation (degrees)						
AC,AN	19.65 (4ZX2)	26.32 (4ZX1)	6.67	20.10	35.85	15.75
BH,AN	51.63 (2ZX2)	66.65 (4ZX2)	15.02	38.55	61.05	22.50
BH,AC	76.79 (2ZX2)	86.72 (4ZX1)	9.93	63.12	86.82	23.70
BS,AN	62.62 (2ZX2)	78.96 (4ZX2)	16.34	61.27	75.24	14.36
BS,AC	81.68 (4ZX2)	89.60 (2ZX1)	7.92	76.34	89.93	13.59
BS,BH	25.71 (CBX1)	29.95 (2ZX2)	4.24	24.92	40.19	15.26

* Regions defined in legend to Figure 2.13.
† The 2Zn molecule 1 isolated trajectory (2ZM1).
Displacements are distances between backbone (N, C_α, C) centroid.
The orientation is the inner angle between the principal axes of inertia of the elements calculated from N, C_α, C atomic positions.
The ranges of RMS differences (in inter-C_α distances) between equivalent elements in the crystal structures are

AN, 0.2–1.0 Å; AC, 0.1–0.3 Å; BH, 0.1–0.4 Å; BS 0.1–0.2 Å

The corresponding RMS variations of the regions in the trajectory are less than 0.2 Å.
Codes in parentheses are from Table 2.2 and indicate the crystal conformer from which the value is derived.

regions of the available secondary structure packing space.

The simulated secondary structure dynamics of the isolated and associated (dimeric) species can be compared with the range of secondary structure packing observed in the crystal conformers. In Table 2.6 the relative displacements and orientations of the selected secondary structure regions found in the crystal conformers of 2Zn, 4Zn and cubic insulin are given. The regions considered typically differ between the conformers by less than 0.3 Å in terms of the RMS difference in their inter-C_α distances (R_d). Larger differences of up to 1.0 Å are found in the comparisons with the AN region of 2Zn molecule 2. The regions are therefore generally very similar and allow a comparison of their relative motions in terms of the rigid body analysis adopted for the analysis of the dynamics.

The crystal data reveal that the displacements between regions vary by 0.1–0.76 Å and orientations by 7–16°. Larger displacement ranges of up to 2 Å are found in the 4Zn molecule 2 species between the A- and B-chains. The range of secondary structure motions in individual simulated trajectories of a given conformer can encompass the range of secondary structure packing found in the different crystal conformers (see Table 2.6 and

Figures 2.14(b) and 2.14(c)). An exception is that the dynamics of molecule 1 conformers does not extend to secondary structure motions such as those observed between the A- and B-chains in the 4Zn molecule 2 crystal conformer and the range of conformations available to the *N*-terminus of the B-chain (see Figure 2.17). This can be contrasted with the study of myoglobin by Elber and Karplus (1987).

INTERCONFORMER RELATIONSHIPS

The aim of this analysis is to examine the overall structural relationships of the insulin conformers both as observed in the crystal and as a result of the molecular dynamics simulations. The approach taken is to examine, in a pairwise manner, the overall similarity of the different conformers and involved crystal, simulation-average and simulation-transient conformers.

Visualizing Conformational Relationships

The simplest and most common measure of the similarity of two protein conformations is an overall RMS difference in some structural parameters (such as atomic positions or interatomic distances). This provides an easily understandable 'one-number' index of similarity. The pairwise comparison of *N* conformations may be arranged in a symmetric matrix of order *N*; the elements of the matrix (some index of similarity) may then be regarded as the 'distance' between conformers in the *N*-dimensional space that their comparisons constitute. If *N* is large, the distribution of conformers in such a multidimensional space is difficult to visualize. Fortunately methods exist which allow for the representation of high dimensional relationships in a lower dimensionality (Gordon, 1981).

Essentially this is done by finding a set of coordinates which generate the distances between conformations. This can be done exactly using points in *N*-1 dimensions; however, commonly points in two (or three) dimensions can provide a useful representation of the full *N*-dimensional space. Thus the relationship between conformers may be visualized by plotting their derived coordinates in (say) two dimensions. As the distance between the coordinates represents the true distance between the conformers, the similarity of the conformers may be judged by their proximity on the two-dimensional plot.

The method for generating the coordinates (principal coordinates analysis or metric scaling) is established in multivariate analysis (Gordon, 1981; Chatfield and Collins, 1980), and involves a diagonalization of the *N*-dimensional distance matrix. The eigenvectors corresponding to the highest two (or three) eigenvalues may be used as the coordinates for the conformers. The method has been used in protein trajectory structural analysis by Levitt (1983) in which can be found a full derivation of the method. The technique is directly analagous to the method of phase space reconstruction in the fractal approach to time series analysis (Broomhead and Jones, 1989). For the work presented here the authors used their own implementation of the method. In this work the index of similarity used is the overall RMS in inter-C_α distances (R_d).

Average Conformers

An average conformer was generated from each simulation over the period 20–100 ps at a resolution of 0.1 ps, by Cartesian coordinate averaging (no removal of net translation or rotation was found necessary). Such an averaging procedure can lead to untypical bonded geometry or poor atomic van der Waals' overlap; the average structures were refined by limited energy minimization. This was done in order to be consistent with the crystallographic refinement procedure, in which stereochemical restraints are applied to enforce good bonded geometry on a model built into averaged data (e.g. Konnert and Hendrickson, 1980). The structural changes as a result of energy minimization are minimal, with RMS displacements in the atomic positions of the backbone (N, C_α, C) atoms of less than 0.2 Å.

In Figure 2.15, a two-dimensional representation of the thirteen-dimensional comparison space of the crystal and simulation-average conformers (omitting the 4Zn molecule 2 species because of their very large dissimilarity) is shown. Although the representation is approximate, it provides an overall view of the interconformer relationships. Immediately obvious from the plot is the aforementioned general trend that the deviations from the initial crystal conformers are greater for the isolated simulations than the dimer simulations (the exception being 4Zn molecule 1). In general a pair of simulation-average conformers are not more similar to one another than their initial crystal conformers are. A trend is observed in which the simulation-average conformers are more similar to the cubic conformers (both crystal and simulated) than they are to their initial crystal conformer.

Transient Conformers

An examination of the conformational relationships of transient structures from the simulations proves revealing of the overall nature of the trajectories. A good example is the cubic dimer simulation where initially the two constituent monomers are identical. In Figure 2.16(a), it is clear that the

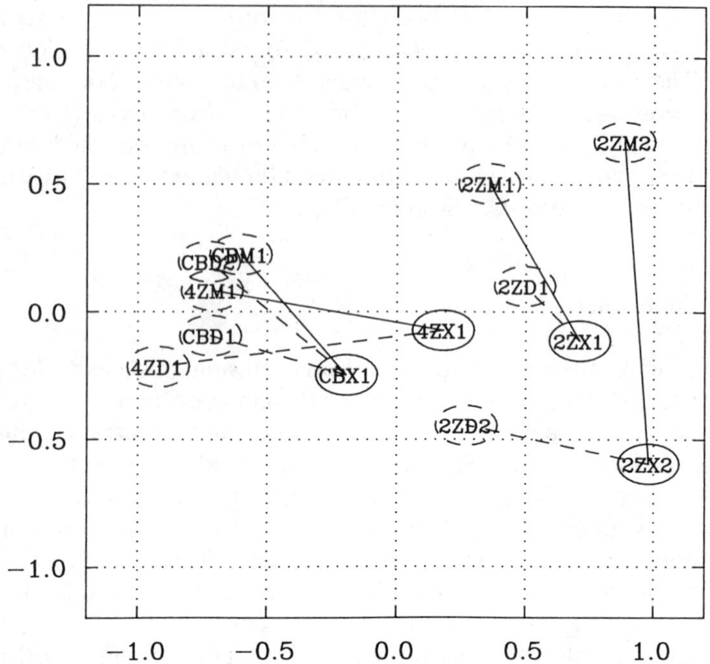

Figure 2.15 Two-dimensional projection of the thirteen-dimensional similarity space constituted by the pairwise comparison of the crystal and simulation-average conformers (excluding the 4Zn molecule 2 crystal and simulated conformers). The distance between conformers on the plot (labelled as given in Table 2.2) represents the value of the RMS difference in inter-C_α distances (R_d) between the conformers. The coordinates of the conformers were generated from a principal coordinates analysis (see text) of the symmetric matrix of pairwise RMS differences. The horizontal axis corresponds to the principal coordinate, the vertical axis to the secondary coordinate. The axes are in Å (1 Å = 0.1 nm). An indication of the error in this representation is given by the RMS difference between the distances derived from the two-dimensional coordinates and those in the full thirteen-dimensional space and is 1.1 Å. Crystal conformers are denoted by full ellipses, simulation-average conformers by broken ellipses. A line is drawn connecting a crystal conformer with its corresponding simulation-average conformers and is broken for the associated simulations and full for the isolated simulations

molecules move away from their crystal conformation and from each other as the simulation proceeds to sample distinct regions of conformational space.

An alternative view of the 2Zn molecule 1 and molecule 2 A-chain conformational differences that were discussed earlier is shown in Figure 2.16(b). In the dimer simulation the A-chain conformation of molecule 1 is very stable and does not drift very far from the crystal conformation (R_d of the simulation-average conformer vs. the crystal conformer of 0.46 Å). In contrast, in the simulation molecule 2 is revealed to move towards the crystal conformation of molecule 1.

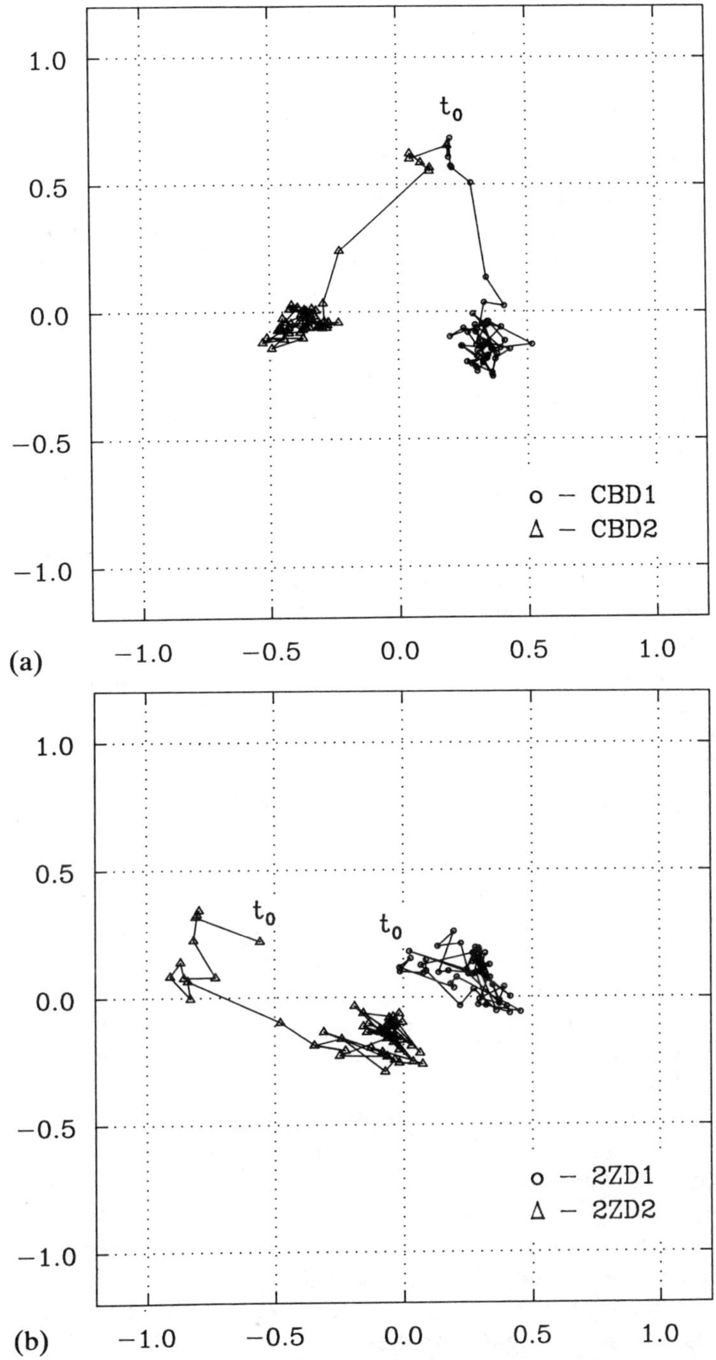

Figure 2.16 Two-dimensional projection of the similarity space constituted by the pairwise comparison of the transient conformers from the simulations. Structures are sampled at 2 ps intervals, over the full 100 ps. The position of the initial (crystal) conformer is indicated by t_0 (units in Å, 1 Å = 0.1 nm). (a) Cubic dimer simulation (RMS error = 0.85 Å). (b) 2Zn dimer simulation (comparisons made on A-chain only) (RMS error = 0.77 Å)

SUMMARY AND CONCLUSIONS

Molecular dynamics simulations of different insulin conformers have been described. The eight simulated trajectories provide a vast amount of data concerning the intrinsic dynamics of the insulin molecule, as a monomer and as a dimer. The trajectories have been examined in a hierarchical manner, from atomic motion and the dynamics of regions of secondary structure, to measures of the overall conformational dynamics of the molecule. The main points emerging from the analysis are now given.

1. Overall structural behaviour in the simulations reveals that prominent changes occur to conformational features in the initial structures that are considered to result from specific crystal packing or medium effects. Examples are the closer approach of chain termini in 2Zn molecule 2 and the contraction of the A- and B-chain separation in 4Zn molecule 2. Structural regions without specific conformation-defining interactions (such as the *N*-terminus of the B-chain) are also found to differ most from their initial conformation as a result of simulation.

2. Atomic fluctuations calculated from simulated trajectories do not reproduce well the trends of fluctuations from crystallographic *B*-factors. A comparison of crystallographic *B*-factors between the different crystal forms reveals a more consistent trend in the B-chain than in the A-chain. The agreement of the simulated fluctuations with experiment is found to be better in the B-chain than in the A-chain.

3. The magnitude and spatial extent of correlated fluctuations in the trajectories is in agreement with results inferred from a study of X-ray thermal diffuse scattering (TDS) data on 2Zn insulin (Caspar *et al.*, 1988). Analysis of the simulations reveals that short-range (4–6 Å) collective behaviour occurs mainly between adjacent residues. Collective motion extending over longer ranges (10–20 Å) is found to coincide with (or to occur at the interface between) defined secondary structure regions (such as the helices).

4. Analysis of the structural integrity of secondary structure regions was performed by monitoring the fluctuation of distances between the atoms they comprise. The results reveal that certain regions have high intrinsic structural stability. Such regions are found to be more extensive around helices, which is thought to be a consequence of the constraints of the helical hydrogen bonding. Such structural invariance, combined with the collective character of their motion, confirms a view of such regions as possessing a relatively rigid character.

5. An analysis of the relative motion of certain structural regions (within the rigid-body limit) allows a mapping of the overall motions of the molecule in terms of secondary structure dynamics. The results reveal that there is a large degree of conformation overlap between the distinct

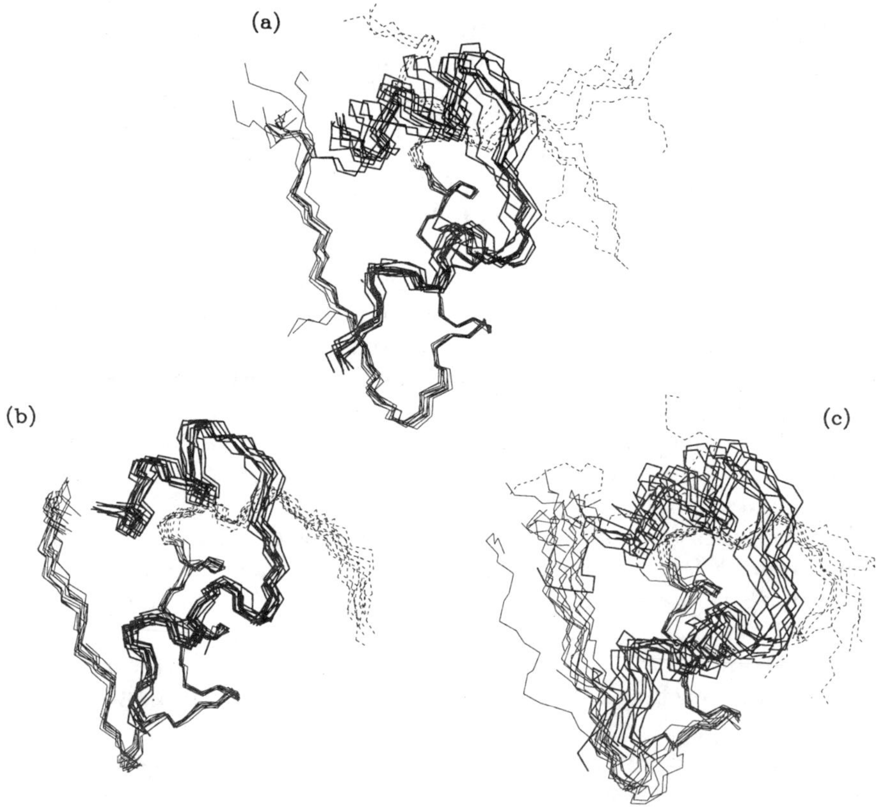

Figure 2.17 Views of the ensemble of insulin conformers found in the crystal and in the molecular dynamics trajectories. The backbone (N, C_α, C) atoms of the structures are shown. The molecules are superposed onto the mainchain (N, C_α, C, O) atoms of the B-helix (B9–B19) of the crystal conformer of 2Zn molecule 1. The A-chain is shown in thick lines, the B-chain in thin lines. The region B1–B8 is shown in dashed lines—this region shows the most variation in the crystal conformers and is found to be highly mobile in the simulations. (a) Eleven distinct native-like insulin crystal conformers. In addition to the five considered in this study, there are the six others, i.e. orthorhombic human insulin molecules 1 and 2, monoclinic human insulin molecules 1 and 2 (there are six independent molecules in this structure), and des-pentapeptide (DPI) beef insulin molecules 1 and 2 (B26–B30 removed). References to these structures may be found in Derewenda *et al.* (1989). Note the three distinct populations of conformations at the *N*-terminus of the B-chain. (b) Eleven transient structures from the 2Zn molecule 1 isolated trajectory sampled at 4 ps intervals over the period 36–80 ps. (c) The eleven distinct simulation-average conformers from the six molecular dynamics simulations considered in the study (see Table 2.2). The conformers are averaged over the period 20–100 ps. For details of the averaging procedure see text. The outlier at the *C*-terminal strand of the B-chain is from the 4Zn molecule 2 conformer in its dimer simulation

conformers in terms of secondary structure packing. The range of motions (displacements, orientations) of a given conformer, in a 100 ps room temperature *in vacuo* trajectory, can encompass the range of packing observed in different crystal forms. However, conformational changes of the magnitude of the difference of 4Zn molecule 2 from the rest of the conformers (which involve an extension of the B-helix at the *N*-terminus and large displacements in the relative position of A- and B-chains) are found not to be sampled in the trajectories (see Figure 2.17).

6. Analysis of the relationships of the simulation-averaged conformers and the crystal conformers proves to be complex. Conformers differ from their initial crystal conformers as a result of simulation, but generally do not become more similar to one another. This can be contrasted to the results of other MD simulations on isolated 2Zn insulin conformers, where molecule 2 was found to become more similar to molecule 1 as a result of simulation (Krüger *et al.*, 1987). Thus for the simulation-average conformers a convergence of the conformational differences is not observed. A trend is found in which simulation-average conformers are more similar to the cubic simulation-average conformers than they are to their own initial crystal conformer.

7. Analysis of the conformational relationships of the transient conformers from different trajectories gives an indication of the extent of conformational space sampled in the simulations. Although the different trajectories remain distinct in conformational space (as measured by RMSDs in interatomic distances), their dynamical properties (as measured by the relative packing of secondary structure regions) can be quite similar.

This simulation study of the conformational dynamics of the insulin molecule has provided an interesting perspective on the wealth of experimental data concerning the conformational variability of insulin. The molecular dynamics simulations were performed *in vacuo* and are of short duration, and therefore do not represent the state of the art in simulation methodology. However, the generation and analysis of trajectories of multiple initial crystal conformers has provided some interesting correlations with experimental data on this highly characterized molecule. Evidence is growing that the ability of the native molecule to undergo a conformational change (from that observed in the crystal) is of functional importance (Dodson *et al.*, 1983; Baker *et al.*, 1988). Although a functionally significant conformational change may be induced only on binding of the molecule to the insulin receptor, it is increasingly apparent that the structure of the insulin molecule has great potential for conformational variability (e.g. Weiss *et al.*, 1989). It is hoped that through simulation studies such as these, based on insulin structures derived from high resolution X-ray diffraction analyses, further insights into the types and mechanisms of conformational changes in proteins may be gained.

ACKNOWLEDGEMENTS

Dr Bruce Tidor is thanked for providing a vectorized version of CHARMM for the Cyber-205. Thanks to the John von Neumann National Supercomputer Center at Princeton where the simulations were performed. Many thanks to Professor Martin Karplus and his group at Harvard for their hospitality and ample provision of advice, encouragement, software and computer time. M.K. is thanked for many useful comments on the manuscript. Thanks to Professor Guy Dodson at York for his unceasing and boundless enthusiasm for structural studies of insulin. The analysis of the trajectories was performed both at Harvard and York D.T.N. and the work at Harvard was supported by a grant from the National Science Foundation (USA). This work forms part of an on-going York–Harvard collaboration on insulin structure and dynamics.

REFERENCES

Adams, M. J., Blundell, T. L., Dodson, E. J., Dodson, G. G., Vijayan, M., Baker, E. N., Harding, M. M., Hodgkin, D. C., Rimmer, R. and Sheet, S. (1969). Structure of rhombohedral 2Zn pig insulin crystals. *Nature*, **224**, 491–496

Alder, B. J. and Wainwright, T. E. (1959). Studies in molecular dynamics. I. General method. *J. Chem. Phys.*, **31**, 459

Anderson, A. and Hermans, J. (1988). Microfolding: conformational probability map for the alanine dipeptide in water from molecular dynamics simulations. *Proteins*, **3**, 262

Åqvist, J., van Gunsteren, W. F., Leijonmarck, M. and Tapia, O. (1985). A molecular dynamics study of the *C*-terminal fragment of the L7/L12 ribosomal protein: secondary structure motion in a 150 ps trajectory. *J. Mol. Biol.*, **183**, 461–477

Artymiuk, P. J., Blake, C. C. F., Grace, D. E. P., Oatley, S. J., Phillips, D. C. and Sternberg, M. J. E. (1979). Crystallographic studies of the dynamic properties of lysozyme. *Nature*, **28**, 563–568

Badger, J. (1986). X-ray crystallographic studies of cubic insulin and of the ribosomal proteins L30 and S5. *D. Phil. Thesis*, University of York

Baker, E. N. and Hubbard, R. E. (1984). Hydrogen bonding in globular proteins. *Prog. Biophys. Mol. Biol.*, **44**, 97–179

Baker, E. N., Blundell, T. L., Cutfield, J. F., Cutfield, S. M., Dodson, E. J., Dodson, G. G., Hodgkin, D. C., Hubbard, R. E., Isaacs, N. W., Reynolds, C. D., Sakabe, K., Sakabe, N. and Vijayan, N. M. (1988). The structure of 2Zn pig insulin at 1.5 Å resolution. *Philos. Trans. R. Soc. London, Sec. B*, **319**, 369–456

Barlow, D. J. and Thornton, J. M. (1988). Helix geometry in proteins. *J. Mol. Biol.*, **201**, 601–619

Bash, P. A., Field, M. J. and Karplus, M. (1987a). Free energy perturbation method for chemical reactions in the condensed phase: a dynamical approach based on a combined quantum and molecular mechanics potential. *J. Am. Chem. Soc.*, **109**, 8092–8094

Bash, P. A., Singh, U. C., Langridge, R. and Kollman, P. A. (1987b). Free energy calculations by computer simulation. *Science*, **236**, 564–568

Bentley, G. A., Dodson, E. J., Dodson, G. G., Hodgkin, D. C. and Mercola, D. A. (1976). The structure of insulin in 4Zn insulin. *Nature*, **261**, 166–168

Bentley, G. A., Dodson, G. G. and Lewitova, A. (1978). Rhombohedral insulin crystal transformation. *J. Mol. Biol.*, **126**, 871–875

Blundell, T. I., Dodson, G. G., Hodgkin, D. C. and Mercola, D. A. (1972). Insulin: the structure in the crystal and its reflection in chemistry and biology. *Adv. Protein Chem.*, **26**, 279–402

Brooks, B. R., Bruccoleri, R. E., Olafson, B., States, D. J., Swaminathan, S. and Karplus,

M. (1983). CHARMM: a program for macromolecular energy, minimization and dynamics calculations. *J. Comput. Chem.*, **4**(2), 187–217

Brooks III, C. L. and Karplus, M. (1986). Theoretical approaches to the solvation of biopolymers. *Methods Enzymol.*, **127**, 369–400

Brooks III, C. L. and Karplus, M. (1989). Solvent effects on protein motion and protein effects on solvent motion: dynamics of the active site region in lysozyme. *J. Mol. Biol.*, **208**, 159–181

Brooks III, C. L., Karplus, M. and Pettit, B. M. (1988). Proteins: a theoretical perspective of structure, dynamics and thermodynamics. *Advances in Chemical Physics*, Vol. LXXI. Wiley, New York

Broomhead, D. S. and Jones, R. (1989). Time series analysis. *Proc. R. Soc. London, Ser. A*, **423**, 103–121

Brünger, A. T. and Karplus, M. (1988). Polar hydrogen placement in proteins: empirical energy placement and neutron diffraction comparison. *Proteins*, **4**, 148–156

Brünger, A. T., Kuriyan, J. and Karplus, M. (1987). Crystallographic *R*-factor refinement by molecular dynamics. *Science*, **235**, 458–460

Case, D. A. and Karplus, M. (1979). Dynamics of ligand binding to heme proteins. *J. Mol. Biol.*, **132**, 343

Caspar, D. L. D., Clarage, J., Salunke, D. M. and Clarage, M. (1988). Liquid-like movements in crystalline insulin. *Nature*, **332**, 659–662

Caves, L. S. D. (1989a). Conformational variability of insulin: a molecular dynamics study. In *Molecular Simulation and Protein Crystallography, Proceedings of a Joint CCP4/CCP5 Study Weekend, January 27–28, 1989*, Daresbury Laboratory, Daresbury

Caves L. S. D. (1989b). Computer simulation studies of native and modified insulins. *D. Phil. Thesis*, University of York

Chatfield, C. and Collins, A. J. (1980). *Introduction to Multivariate Analysis.* Chapman and Hall, London

Chothia, C., Dodson, G. G., Hodgkin, D. C. and Lesk, A. (1983). Transmission of conformational change in insulin. *Nature*, **302**, 500–505

Cutfield, J. F., Cutfield, S. M., Dodson, E. J., Dodson, G. G., Reynolds, C. D. and Valley, D. (1981). Similarities and differences in the crystal structures of insulin. In Dodson, G. G., Glusker, J. P. and Sayre, D. (Eds.), *Structural Studies on Molecules of Biological Interest.* Oxford University Press, Oxford, pp. 527–546

Derewenda, U., Derewenda, Z. and Dodson, G. G. (1988). Manuscript in preparation

Derewenda, U., Derewenda, Z., Dodson, G. G., Hubbard, R. E. and Korber, F. (1989). Molecular structure of insulin: the insulin monomer and its assembly. *Br. Med. Bull.*, **45**(1), 4–18

Dodson, E. J., Dodson, G. G., Lewitova, A. and Sabesan, M. (1978). Zinc-free cubic pig insulin: crystallization and structure determination. *J. Mol. Biol.*, **125**, 387–396

Dodson, E. J., Dodson, G. G. and Hodgkin, D. C. (1980a). The conformations observed in the *N*-terminal A-chain residues of insulin. In Ananchenko, S. N. (Ed.), *Frontiers of Bioorganic Chemistry and Molecular Biology*. Pergamon, Oxford, pp. 245–250

Dodson, E. J., Dodson, G. G., Reynolds, C. D. and Valley, D. (1980b). Comparison of 2Zn and 4Zn insulin crystal structures. In Brandenburg, D. and Wollmer, A. (Eds), *Insulin, Chemistry, Structure and Function of Insulin and Related Hormones*. de Gruyter, New York, pp. 9–16

Dodson, G. G., Hubbard, R. E. and Reynolds, C. D. (1983). Insulin's structural variation and its relation to activity. *Biopolymers*, **22**, 281–292

Doucet, J. and Benoit, J. P. (1987). Molecular dynamics studied by X-ray diffuse scattering from lysozyme crystals. *Nature*, **325**, 643–646

Elber, R. and Karplus, M. (1987). Multiple conformational states of proteins: a molecular dynamics analysis of myoglobin. *Science*, **235**, 318–321

Frank, B. H., Pekar, A. H. and Veros, A. J. (1972). Insulin and proinsulin conformation in solution. *Diabetes*, **21**, 486–491

Gelin, B. R. (1976). Application of empirical energy functions to conformational problems in biochemical systems. *Ph.D. Thesis*, Harvard

Goldman, T. and Carpenter, F. (1976). Zinc-binding, circular dichroism and equilibrium sedimentation studies of insulin (bovine) and several derivatives. *Biochemistry*, **13**, 486–491

Gordon, A. D. (1981). *Classification.* Chapman and Hall, London van Gunsteren, W. F. and Berendsen, H. J. C. (1977). Algorithms for molecular dynamics and constraint dynamics. *Mol. Phys.*, **34**(5), 1311–1327

Havel, T. F., Kuntz, I. D. and Crippen, G. M. (1983). The theory and practice of distance geometry. *Bull. Math. Biol.*, **45**, 665

Hermans, J. (1985). Potential functions for the simulation of large molecules. In Hermans, J. (Ed.), *Molecular Dynamics and Protein Structure.* Polycrystal Book Services, Chapel Hill, pp. 31–34

Kaarlsholm, N. C., Hui, K. and Dunn, M. F. (1989). Comparison of solution structural flexibility and zinc binding domains for insulin, proinsulin and miniproinsulin. *Biochemistry*, **28**(10), 4427–4435

Kabsch, W. (1976). A solution to the best rotation to relate two sets of vectors. *Acta Crystallogr. A*, **32**, 922–923

Kaptein, R., Zuiderweg, E. R. P., Scheek, R. M., Boelens, R. and van Gunsteren, W. F. (1985). A protein structure from nuclear magnetic resonance data: lac repressor headpiece. *J. Mol. Biol.*, **182**, 179–182

Konnert, J. H. and Hendrickson, W. A. (1980). A restrained parameter thermal factor refinement procedure. *Acta Crystallogr. A*, **36**, 344

Krüger, P., Straßburger, W., Wollmer, A., van Gunsteren, W. F. and Dodson, G. G. (1987). The simulated dynamics of the insulin monomer and their relationship to the molecule's structure. *Eur. Biophys. J.*, **14**, 449–459

Lesk, A. M. and Chothia, C. (1980). How different amino-acid sequences determine similar protein structures. Structure and evolutionary dynamics of the globins. *J. Mol. Biol.*, **136**, 225–270

Levitt, M. (1983). Molecular dynamics of native protein: II. Analysis of trajectories. *J. Mol. Biol.*, **168**, 621–657

Levitt, M. and Sharon, R. (1988). Accurate simulation of protein dynamics in solution. *Proc. Natl. Acad. Sci. USA*, **85**, 7557–7562

McCammon, J. A. (1984). Protein dynamics. *Rep. Prog. Phys.*, **47**, 1–46

McCammon, J. A. and Harvey, S. C. (1987). *Dynamics of Proteins and Nucleic Acids.* Cambridge University Press, Cambridge

McCammon, J. A., Gelin, B. R. and Karplus, M. (1977). Dynamics of folded proteins. *Nature*, **267**, 585–590

McCammon, J. A., Lee, C. Y. and Northrup, S. H. (1983). Sidechain rotational isomerization in proteins: a mechanism involving gating and transient packing defects. *J. Am. Chem. Soc.*, **105**, 2232

Peking Insulin Structure Group (1971). Insulin's crystal structure at 2.5 Å resolution. *Peking Rev.*, **40**, 11

Peking Insulin Structure Group (1974). Studies on the mechanisms of insulin action. *Sci. Sin.*, **17**, 752–778

Post, C. B., Brooks, B. R., Karplus, M., Dobson, C. M., Artymiuk, P. J., Cheetham, J. C. and Phillips, D. C. (1986). Molecular dynamics simulations of native and substrate bound lysozyme. A study of average conformers and atomic displacements. *J. Mol. Biol.*, **190**, 455–479

Post, C. B., Dobson, C. M. and Karplus, M. (1989). A molecular dynamics analysis of protein structural elements. *Proteins*, **5**(4), 337–354

Powell, M. J. D. (1977). Restart procedures for conjugate gradient methods. *Math. Program.*, **12**, 241–245

Ramesh, V. and Bradbury, J. H. (1986). ^{1}H NMR studies of insulin: reversible transformation of 2Zn→4Zn insulin hexamer. *Int. J. Pept.*, **28**(2) 146–153

Renscheidt, H., Strassburger, W., Glattner, U., Wollmer, A. and Mercola, D. (1984). A solution equivalent of the 2Zn→4Zn transition of insulin in the crystal. *Eur. J. Biochem.*, **142**, 7–14

Ringe, D. and Petsko, G. A. (1985). Mapping protein dynamics by X-ray diffraction. *Prog. Biophys. Mol. Biol.*, **45**, 197–235

Ryckaert, J. P., Ciccotti, G. and Berendsen, H. J. C. (1977). Numerical integration of the Cartesian equations of motion with constraints: molecular dynamics of *n*-alkanes. *J. Comput. Phys.*, **23**, 327–341

Swaminathan, S., Ichiye, T., van Gunsteren, W. F. and Karplus, M. (1982). Time dependence of atomic fluctuations in proteins: analysis of local and collective motions in the bovine pancreatic trypsin inhibitor. *Biochemistry*, **21**, 5230–5241

Verlet, L. (1967). Computer 'experiments' on classical fluids. I. Thermodynamic properties of Lennard-Jones molecules. *Phys. Rev.*, **159**, 98

Weiner, S. J., Kollman, P. A., Case, D. A., Singh, U. C., Ghio, C., Alagona, G., Profeta, S. and Weiner, P. (1984). A new force-field for the molecular mechanical representation of nucleic acids of proteins. *J. Am. Chem. Soc.*, **106**, 765–784

Weiss, M. A., Nguyen, D. T., Khait, I., Inouye, K., Frank, B. H., Beckage, M., O'Shea, E., Shoelson, S. E., Karplus, M. and Neuringer, L. J. (1989). Two dimensional NMR and photo-CIDNP studies of the insulin monomer. Assignment of aromatic resonances with applications to protein folding, structure and dynamics. *Biochemistry*, **28**, 9855–9873

Willis, B. T. M. and Pryor, A. W. (1975). *Thermal Vibrations in Crystallography*. Cambridge University Press, Cambridge

Wodak, S. J., Alerd, P., Delhause, P. and Renneboog-Squilbin, C. (1984). Simulation of conformational changes in 2Zn insulin. *J. Mol. Biol.*, **181**, 317–322

Wollmer, A., Fleischhauer, J., Strassburger, W., Thiele, H., Brandenburg, D., Dodson, G. and Mercola, D. (1977). Sidechain mobility and the calculation of tyrosyl circular dichroism of proteins: implications of a test with insulin and des-B1 phenylalanine insulin. *Biophys. J.*, **20**, 233–243

Wong, C. F. and McCammon, J. A. (1986). Dynamics and the design of enzymes and inhibitors. *J. Am. Chem. Soc.*, **108**, 3830–3832

3
Molecular Dynamics Simulations of Carbohydrates

Jutta Köhler

INTRODUCTION

In the past three decades carbohydrate chemistry has been neglected when compared with the rapidly developing chemistry and biochemistry of proteins and nucleic acids. Even so, some basic characteristics have been studied, such as the anomeric effect in sugar residues (Wolfe *et al.*, 1979), chirality (Ruch, 1972) and the conformational properties of sugars and their glycosidic linkages (Lipkind *et al.*, 1984). Recently the general interest in this field (Kennedy, 1988) has grown, resulting in studies about the conformation and kinetics of polysaccharides (Goodall and Norton, 1987), the three-dimensional structure of starch (Imberty and Perez, 1988), furanose ring flexibility (Pechenaya and Serikov, 1988), *gauche* interactions (Wiberg and Murcko, 1988) and hydrogen bonding in oligosaccharides (Davies and Christofides, 1987). Other studies have appeared on the chemical modification of polysaccharides (Aspinall, 1987) and on industrial genetic engineering and its application (Yalpani, 1988).

Oligosaccharides and related molecules such as glycopeptides and glycoproteins play an important role in human biochemistry. For example, the human blood group systems contain oligosaccharide determinants linked to protein (Lemieux, 1978). In this respect the carbohydrates are correlated with the immune system and therefore are of greater medical interest (Skerra and Plueckthun, 1988; Plueckthun and Skerra, 1989). It has already become possible to produce antibodies with exchanged amino acids in bacteria, so the design of glycoprotein molecules with desired functions might be possible in the future as well (Karplus, 1987).

The cyclodextrins are cyclic oligosaccharides and have been the subject of a broad range of investigations (Saenger, 1980; Szejtli, 1984). Compared with linear saccharides they display the ability of forming host–guest complexes (Atwood *et al.*, 1984) and serve as models for catalytic enzymes. Cyclodextrins allow the study of photochemical effects (Ramamurthy and Eaton, 1988), chiral recognition (Hamilton and Chen, 1988), solvent–salt effects on binding constants, hydrophobic–hydrophilic binding contributions (Schneider *et al.*, 1988), and hydrogen bonding patterns (Lesyng and Saenger, 1981). The influence of solvent and temperature on the optical rotation behaviour of model carbohydrates has been studied by Rees and Thom (1977). Thus carbohydrate chemistry is linked to many subjects including physics, chemistry, biochemistry, biology and medicine with applications even in the field of agriculture (Saenger, 1980).

Carbohydrates have also been studied using theoretical techniques. Since (except for the cyclodextrins and a few fortunate examples) oligosaccharides are difficult to crystallize, there is a need to use molecular modelling techniques in order to explain the observed structures, properties and kinetics. What is happening, for example, in extended amylose chains near their minimum energy conformation (Neszmelyi and Hollo, 1989)? How flexible are these structures and what dynamical behaviour do they have in different environments? There are also more difficult questions: which inclusion complexes can be formed (Linert *et al.*, 1989) and what reactions can they undergo or catalyse? The development of theoretical methods to describe and analyse such questions will be outlined in the following section.

NOTES ON THEORETICAL METHODS

Calculations on biomolecular systems such as molecular modelling (Rees and Scott, 1971), computer-aided molecular design (Kollman, 1985; Berendsen, 1988; Richards, 1989) and visualization using computer graphics (Burgen *et al.*, 1986) are regularly used to understand large three-dimensional structures and their correlation with biological function. Complex formation of molecules is of special interest for analysing binding properties (e.g. docking calculations on ferrocene complexation with cyclodextrins have been performed by Menger and Sherrod, 1988).

Current methods have their roots in the early 1960s, when the conformational analysis of macromolecules started to become of more general interest (Ramachandran and Sasisekharan, 1968; Brant *et al.*, 1969). Anderson *et al.* (1969) have used model building and X-ray diffraction studies to determine the double helical structures of polysaccharides using crystal structure data as an initial set of coordinates followed by computational sampling of new structures by rotation around selected covalent bonds. The details of these so-called hard-sphere calculations are described in Rees and Skerrett (1968) and Rees and Smith (1975). This approach has also been used for carbohydrate conformations in the analysis of bacterial polysaccharide structures and linkages (Bock *et al.*, 1984; Talashek and Brant, 1987).

Another approach to the calculation of static structures of larger systems is the molecular mechanics method (Allinger and Sprague, 1972; Melberg and Rasmussen, 1978, 1979). Originally it used many different terms for each atom type but was simplified later to a more general force field with terms for bond lengths, angles and torsion angles, and electrostatic and van der Waals terms. This method could be applied to much larger molecular systems than quantum mechanical methods (Allinger and Graham, 1973). A number of algorithms are available such as the steepest descent algorithm or the conjugate gradient method (Fletcher and Reeves, 1964). The method has been applied to the study of serogroup polysaccharides (Lindon *et al.*, 1984).

The Monte Carlo method (Rees and Smith, 1975) is based on similar force fields, but in addition the large multidimensional conformational energy hypersurface of all possible combinations of bond lengths, angles and torsional angles is scanned statistically. Using an energy criterion to accept or discard generated structures, a sampling is performed to give an overview of the most probable conformational structures. There remains still a general problem to distinguish a local minimum from the global minimum conformation.

Ab initio SCF energy calculations have been used to determine the structures of several carbohydrates (Jeffrey *et al.*, 1974; Darsey *et al.*, 1988). The PCILO method (Diner *et al.*, 1969) has been applied to carbohydrates (Struempel and Luger, 1988). The X–α method (Case, 1982) can be parametrized for metal atoms so that calculations on sugar–metal complexes are possible (for example, the studies on sugar–metal interactions by Tajmir-Riahi, 1988).

Several molecular dynamics (MD) packages are available in which the movement of atoms on a picosecond timescale is calculated by integration of Newton's equations of motion. Thus the dynamical behaviour of large biomolecules can be calculated, their flexibilities analysed and displayed. Crystal structures may be simulated and compared with X-ray data (van Gunsteren *et al.*, 1983) or crystallographic refinement methods may be combined with molecular dynamics (Fujinaga *et al.*, 1989). There is also the possibility of analysing the dynamical behaviour of molecules in solution (Allen and Tildesley, 1987). The development of flexible or polarizable water models is of high interest for improving on the MD results in solution and explaining cooperative, non-additive and ionic effects (Stillinger and David, 1978; Berendsen *et al.*, 1987; Ahlstrom *et al.*, 1989). Applications of the molecular dynamics method to study carbohydrates are described in a study on α-D-glucose (Brady, 1986) and on β-D-glucopyranose (Brady, 1987).

MOLECULAR DYNAMICS CALCULATIONS ON CYCLODEXTRINS

We have applied the molecular dynamics method to a well-defined class of oligosaccharides, the cyclodextrins. These molecules were selected because of their interesting properties and their use in practical applications (Saenger, 1980) and because of the availability of a large body of experimental data at the atomic level, obtained from X-ray and neutron diffraction experiments.

Cyclodextrins are torus-shaped oligosaccharides, consisting of six, seven or eight (α-, β-, or γ-cyclodextrin) glucose units that are covalently linked to each other by α- (1–4) glycosidic bonds. The cyclodextrin molecules have a central cavity of about 0.57–0.95 nm and form inclusion compounds (Cramer *et al.*, 1967) with molecules that fit into these cavities. The guest molecules may vary from hydrophilic to hydrophobic in character, the only condition for inclusion being that they are small enough to fit spatially into the cavity. Therefore, cyclodextrins and their inclusion compounds have found applications as molecular capsules in the pharmaceutical (Szejtli, 1977), agricultural (Szejtli, 1983) and food industries. They are also promising materials in gel-inclusion chromatography (Szejtli *et al.*, 1978) and even soluble cyclodextrin polymers have been produced (Szejtli, 1984). Owing to their microheterogeneous properties brought about by the hydrophobic cavity and the hydrophilic rims lined with O–H groups, the cyclodextrins exert diverse catalytic activities which make them good enzyme models (Gelb *et al.*, 1979).

In the crystalline state cyclodextrins form extended hydrogen-bonded networks which involve their hydroxyl groups and the crystal water molecules (Hingerty *et al.*, 1984). The interactions in these hydrogen-bonded networks are rather sensitive to weak long-range forces such as polarization of the covalent O—H bond. They lead to an energetically favourable association of molecules, as indicated by the extension of hydrogen-bonded water molecules and hydroxyl groups to infinite chains and cycles in the cyclodextrin crystals (Saenger, 1979), which is due to a 'cooperative effect' (Barnes *et al.*, 1979). This phenomenon has been investigated by *ab initio* quantum mechanical methods followed by many-body analysis from which it was shown that non-additive contributions account for about 30% of the total interaction energy for a water tetramer with S_4 symmetry (Koehler, 1985; Koehler and Saenger, 1987; Koehler *et al.*, 1987a).

A dynamical equilibrium between two mutually exclusive orientations of hydrogen bonds as shown below is called a flip–flop hydrogen bond:

$$\text{O—H}\cdots\text{O} \rightleftharpoons \text{O}\cdots\text{H—O}$$

Its occurrence was detected experimentally in β-cyclodextrin crystals (Saenger *et al.*, 1982). The neutron difference density maps of β-cyclodextrin crystals at 293 K (Betzel *et al.*, 1984) display many examples of two partially occupied hydrogen sites at a distance of about 0.1 nm between two oxygen sites. This hydrogen distance is too short to allow two hydrogen atoms to be in that area at the same time. The neutron diffraction crystal structure of β-cyclodextrin at 120 K (Zabel *et al.*, 1986) only displays four hydrogen bond structures with partially occupied hydrogen atom sites. With the additional information of the existence of a phase transition at 227 K, it was concluded that the disordered hydrogen bonds in the crystal structure of β-cyclodextrin at 293 K are an averaged structure of a dynamical equilibrium rather than of a statistical distribution in the crystal. The mechanism of this dynamical equilibrium is believed to be either a rotational or a tunnelling mechanism (Koehler, 1985).

Three-centre hydrogen bonds (Kollman and Allen, 1972; Jeffrey and Maluszynska, 1986) occur to about 25% in crystal structures of small biological molecules (Scheme 3.1).

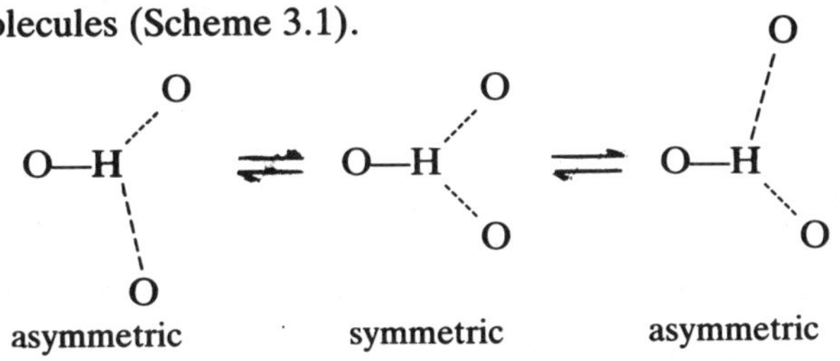

Scheme 3.1

In solution, they could occur as an intermediate state of two different hydrogen bond configurations, and their behaviour in solution structures is of importance for the dynamical interactions between water molecules and atoms at the surface of biological molecules (Saenger *et al.*, 1986).

The cyclodextrins with their O—H---O hydrogen bonding flip–flop dynamics and three-centre hydrogen bond configuration, as observed in neutron diffraction studies in which hydrogen positions were obtained with high accuracy, are a very good test case for the potential force field that is applied in molecular dynamics studies.

The Simulated Cyclodextrin Systems

Molecular dynamics simulations can be applied to molecular systems with up to 10 000 atoms over about 100 ps on currently available computers. Experimentally obtained neutron diffraction structures of α-cyclodextrin hexahydrate at 293 K and β-cyclodextrin undecahydrate at 293 K and 120 K have been used as starting structures to simulate these crystals and additionally α-cyclodextrin in solution has been simulated at 293 K. The cyclodextrin molecules therefore can be analysed using this theoretical method and the results compared with the experimentally obtained data. We are interested in addressing whether the experimentally observed properties of the molecules are well described by the simulation and what additional properties can be obtained from the simulations.

MOLECULAR DYNAMICS SIMULATION OF CRYSTALLINE α-CYCLODEXTRIN HEXAHYDRATE AT 293 K

α-Cyclodextrins crystallize in several forms with different numbers of water molecules inside and outside the cavity. Glucose unit 5 (see Figure 3.1) is rotated in form I (Manor and Saenger, 1974; Klar *et al.*, 1980) and form II (Lindner and Saenger, 1982) to diminish the cavity so that the enclosed water molecules are held tightly. In form III the α-cyclodextrin adopts a 'round' conformation as observed in complexes with other guest molecules (Chacko and Saenger, 1981).

Method and Computational Details

A 15 ps molecular dynamics simulation was performed of the crystal structure of α-cyclodextrin hexahydrate form I at 293 K (Koehler *et al.*, 1987b) using the GROMOS computer program. In this simulation all 1632 solute atoms and water molecules are free to move and all atoms are treated explicitly except for hydrogens attached to carbons. The starting structure (Figure 3.1) for the simulation is taken from the neutron diffraction study by Klar *et al.* (1980). The $C_{36}H_{60}O_{30}\cdot 6H_2O$ complex crystallizes at room temperature in the orthorhombic space group $P2_12_12_1$

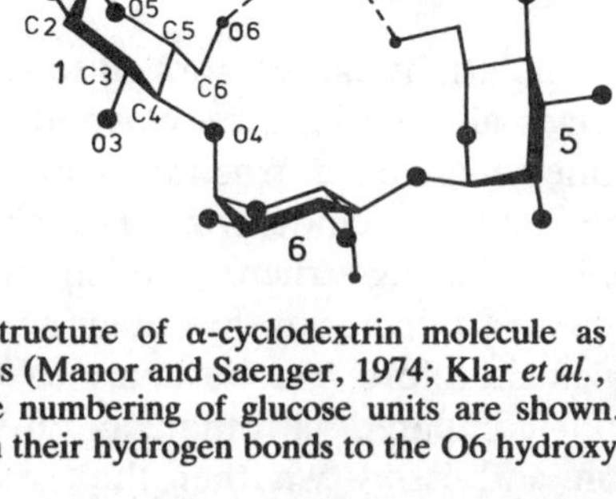

Figure 3.1 Schematic structure of α-cyclodextrin molecule as derived from X-ray and neutron diffraction studies (Manor and Saenger, 1974; Klar *et al.*, 1980). Atom names in the first glucose unit and the numbering of glucose units are shown. The two included water molecules are shown with their hydrogen bonds to the O6 hydroxyl groups of glucose units 1 and 5

with unit cell dimensions $a = 1.4858$ nm, $b = 3.4038$ nm and $c = 0.9529$ nm. The structure was refined to an R-value of 12% for 2808 neutron data and to 3.7% for 4268 X-ray data (Manor and Saenger, 1974). The hydroxyl O61 position is found to be twofold disordered, with the O61A site 92% occupied and the O61B site 8%; all other atom sites are fully occupied. The hydrogen atom HO61B attached to the low occupancy site cannot be located from the neutron diffraction data, but the hydrogen atom attached to the high occupancy site, HO61A, is clearly found. O61A site coordinates are used in the MD simulation. The two water molecule sites within the α-cyclodextrin torus (see Figure 3.1) are denoted by OWA and OWB, the others by OW1–OW4. The computational box is orthogonal and consists of four unit cells each containing 4 α-cyclodextrin and 24 water molecules, with dimensions $2a$, b, $2c$.

The potential function of GROMOS is used with a dielectric constant $\varepsilon_r = 1$. When evaluating the summation of the non-bonded terms, a cut-off radius $R_c = 0.8$ nm was applied, beyond which no interactions were included. Even with this value for the cut-off radius, the summation over non-bonded interactions is typically over 10^5 atom pairs. To reduce computing costs, the list of non-bonded atom pairs is kept constant over 10 MD time steps of $\Delta t = 2$ fs. The cut-off radius is applied to centres of geometry of neutral atom groups and to the oxygen atoms of the water molecules, in order to avoid the breaking of the charge neutrality of a group or of a water molecule which would be possible if an atom–atom cut-off were applied. All bond lengths are kept rigid during the simulation by using the SHAKE method (van Gunsteren and Karplus, 1982). The water molecules are modelled by a simple rigid three-point charge model (SPC; Berendsen *et al.*, 1981; Postma, 1985). Before starting the MD simulation the initial configuration is energy minimized in order to release possible strain. Initial velocities for the atoms are taken from a Maxwellian distribution at 293 K, independently for each of the molecules. The system is weakly coupled to a thermal bath of $T_0 = 293$ K (Berendsen *et al.*, 1984) using a temperature relaxation time $\tau_T = 0.1$ ps. This value makes the temperature coupling weak enough to avoid any significant effect on the atomic properties of the system. Periodic boundary conditions corresponding to the crystal translational symmetry ($2a$, b, $2c$) are applied. The MD run covered a time span of 15 ps. After a few picoseconds the total potential energy remains on a constant level and the root mean-square positional differences between the actual structure and the initial structure also reach a constant value. Therefore, the equilibrium period was chosen to be 5 ps. For the final 10 ps, coordinates are stored every 0.01 ps to be used for analysis. The molecules in the 16 asymmetric units moved independently, which means that statistical averaging could be performed over the time period 5–15 ps and over the 16 asymmetric units. This average is denoted as ⟨M1–16; 5–15 ps⟩.

α-Cyclodextrin Structure and Dynamics

The structure of α-cyclodextrin hexahydrate as derived from the MD simulation is obtained by averaging over the 16 molecules during the time period from 5 to 15 ps. This averaging yields the atomic root mean-square positional fluctuations:

$$\mathrm{RMS}_{\mathrm{pos.fluc.}} = \left[\sum_{\substack{i=1 \\ x,y,z}}^{N} \langle (x_i^{\mathrm{MD}} - \langle x_i^{\mathrm{MD}} \rangle)^2 \rangle \Big/ N \right]^{1/2} \tag{3.1}$$

where the summation i runs over a specified set of N atoms and over the three Cartesian components. Averaging over corresponding atoms in different asymmetric units and over time is denoted by ⟨...⟩.

The average structure is compared with the experimental one in terms of the root mean-square difference of the atomic positions

$$\mathrm{RMS}_{\mathrm{pos.diff.}} = \left[\sum_{\substack{i=1 \\ x,y,z}}^{N} \langle (x_i^{\mathrm{exp}} - \langle x_i^{\mathrm{MD}} \rangle)^2 \rangle \Big/ N \right]^{1/2} \tag{3.2}$$

If not stated otherwise, averaging means taking the root mean-square average. The coordinates of non-hydrogen atoms of the sugar differ from experiment by about 0.018 nm (lowest 0.011 nm, highest 0.023 nm), whereas for the three hydroxyl hydrogen atoms the average is around 0.040 nm (0.029–0.054 nm). The overall averaged RMS deviation for all atoms is 0.025 nm. Comparison of an arbitrarily chosen molecule with the experimental data leads to larger RMS values with an overall average deviation of 0.039 nm for all atoms. The change in the geometry of each molecule during the MD simulation can be different; for example, the difference between the two molecules M1 and M5 is of the same order of magnitude as the deviation of each of them from the experimental structure. Averaging over the 16 molecules in the computational box does improve the agreement with experiment significantly. Averaging over 5 ps instead of 10 ps does not significantly change the resulting deviations and fluctuations. When comparing the structures of two molecules it is common practice to fit them as well as possible to each other by superimposing the centres of mass of both molecules and subsequently performing a least-squares rotational fit around the shared centre of mass. When performing a translational fit the agreement with the experimental structure improves marginally from 0.025 nm to 0.024 nm. When performing a rotational fit to the C3 carbon atom positions the agreement is 0.023 nm. Throughout this study no fit procedure is applied.

Crystallographic refinement techniques yield atomic B-factors which give an indication of atomic thermal vibrations in the crystal lattice. The isotropic *B*-factor is related to the atomic RMS positional fluctuation by

$$\mathrm{RMS}_{\mathrm{pos.fluc.}} = (3B_{\mathrm{iso}}/8\pi^2)^{1/2} \tag{3.3}$$

The experimental and MD values for isotropic RMS positional fluctuations are similar, the MD values being about 0.01 nm larger than the experimental ones, but the relative sizes for the various atoms were well reproduced in the simulation. The MD values are significantly larger than the crystallographic ones for atoms at the end of a covalent chain, such as HO2, HO3 and HO6. This may partly be due to the finding that crystallographic refinement procedures tend to underestimate the *B*-factors of mobile side chain atoms (Kuriyan *et al.*, 1986). The experimental fluctuations of the main chain atoms are well reproduced by the simulation. For the eight 'side chain' atoms HO21, HO24, HO25, HO26, HO36, O61, HO61 and HO63 the mobilities were much larger than reflected in the crystallographic temperature factors. The deviation of the MD structure from the experimental one is smaller than the experimental fluctuation as derived from Equation (3.3). This means that the MD simulation reproduced the atomic positions in the crystal lattice within their experimentally determined accuracy.

From crystallographic data, anistropic *B*-factors are available for non-hydrogen atoms. The ratio between the shortest and longest axes of the thermal motion ellipsoids averaged over the six glucose units has been calculated. For the individual atoms the experimental values range from 0.54 to 0.90 and the MD values from 0.40 to 0.84. The simulation shows more anisotropy in the atomic motions and also a larger variation in anisotropy when comparing atoms in different glucose units. As with the isotropic *B*-factors, the anisotropic refinement procedures tend to underestimate the anisotropy of mobile atoms. This may be checked by applying the crystallographic anisotropy refinement procedure to the diffraction intensities as they can be directly calculated from the simulated trajectories of the system.

Another possibility for assessing the agreement between the experimental and simulated structure is to compare the various bond angles and torsion angles. All bond angles of the α-cyclodextrin glucose ring and O–H groups are well reproduced within 0.9° to 5.6°. The exception is the angle C2–O2–HO2, which has an experimental value of 108.5°, compared with an MD value of 114.3°. The MD bond angle fluctuations range from 3.2° to 4.7°. The glucose ring torsion angles show a slightly larger flexibility; their MD fluctuations range from 5.6° to 6.7°. As to the differences between experiment and MD simulation, torsion angles around C4 (C2–C3–C4–C5 and C3–C4–C5–O5) are about 4.6° smaller in MD than in experiment, indicating a general flattening at the glycosidic link atom.

The difference between MD and experimental glycosidic bond torsion angles are within 5–6°. The largest deviation of 12.4° is observed for the angle C36–C46–O46–C15 (experimental 170.9°, MD 158.5°). The large experimental value for this angle reflects the twisted orientation of glucose unit 5 when compared with the other units, where C3–C4–O4–C1 angles are 116.8–134.7°. We note that the relative variation of the values for the different glucose units is also observed in MD. The glycosidic bond torsion angles are slightly more flexible than the ring torsion angles. Glucose torsion angles involving O2, O3 and O6, but not hydrogen atoms, are reproduced within an average deviation of 4°, where deviations including the O2 and O3 atoms range from 2° to 4° and deviations for O6 atoms are 6° on average. The latter contains two major exceptions. One was O6 of the twisted glucose unit 5 with a difference of 12.9° between the MD and the experimental value. O65 made a hydrogen bond to an internal water molecule OWA, whereas the other O6 atoms are connected to other water molecules or intermolecularly, to hydroxyl groups. The other exception is O61 which forms a strong (78.7%) intermolecular hydrogen bond to O63 of molecule 11 and a weak (2.2%) intramolecular bond to O65, which causes the large fluctuation of 29°. This fluctuation reflects the experimentally observed disorder of this group (the only one in the crystal structure), which is 92% occupied in O61A and 8% in O61B positions which correspond to O5–C5–C6–O6 torsion angles in the ranges +*gauche* (O61A) and −*gauche* (O61B). The average fluctuation of the torsion angles with O6 is about 15°; that is, twice the value of 7–8° for angles with O2 or O3.

The sinoidal potentials of the C2–O2, C3–O3, C5–C6 torsion angles are

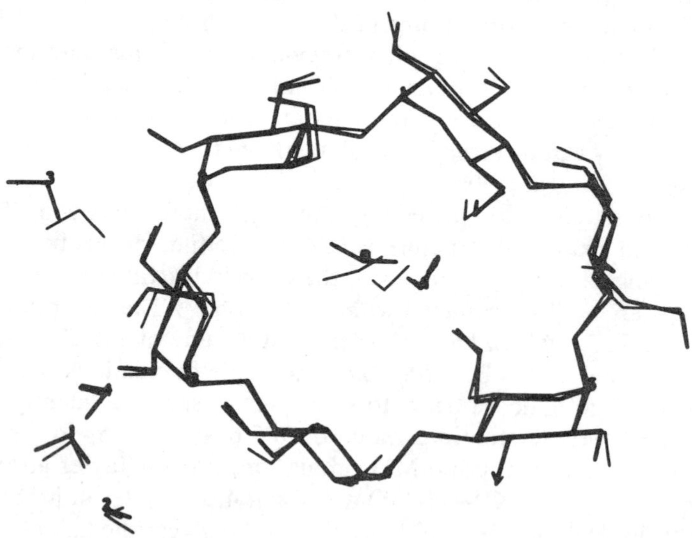

Figure 3.2 α-Cyclodextrin·$6H_2O$ form I at 293 K. Thin lines, neutron diffraction structure; thick lines, averaged MD structure

defined with a threefold symmetry so that the rotational potential showed three barriers and three minima. For the torsion angle C4–C5–C6–O6 of glucose unit 1 eight transitions from one to the other minimum are found during the period from 5 to 15 ps. This result agrees with the experimental finding that O61 is twofold disordered and forms several hydrogen bonds to O63 (intermolecularly), to O65, O56 (intramolecularly), and also to the external water W3. The torsion angles involving atoms H2, H3 and H6 showed many 120° transitions: 166, 66 and 31 respectively for all 16 α-cyclodextrin molecules. This means an average of 1.7, 0.7 and 0.3 transitions per torsion angle per 10 ps. The rotational freedom of these torsion angles is also demonstrated by their large fluctuations, which were 163.5°, 27.7° and 159.4° on average. Very large fluctuations for the H6 torsion angles in glucose units 1 and 5 are related to the occurrence of the twist of unit 5 which also allows for large fluctuations of H2 and H3 torsion angles in unit 6. The large fluctuation of the H2 torsion angle in unit 4 suggests that this hydroxyl group is involved in several hydrogen bonds; it was bound to O35, to O54, and to a water WB. An overall impression of the agreement between the simulated and the experimental α-cyclodextrin structures can be obtained from Figure 3.2.

Water Molecules

Of the six water molecules in the asymmetric unit, the experimentally observed positions of W1, W2, W4 and WB are reproduced by the MD simulation within their experimental accuracy. Water molecules W3 and WA, which showed the largest positional fluctuations in the simulations, are shifted from their experimental positions by 0.11 and 0.12 nm, when averaged over the 16 asymmetric units. All W1, W2, W4 and WB waters in the 16 asymmetric units have positional differences of less than 0.08 nm. The largest deviation for waters WA and W3 is about 0.2 nm, except for the two asymmetric units where water W3 drifted away from its experimental position by 0.36 nm and 0.44 nm. These shifts in water positions might be associated with the formation of dynamically disordered O—($\frac{1}{2}$H)----($\frac{1}{2}$H)—O hydrogen bonds of the flip–flop type which were observed in the MD simulation (2.2% and 5.1% for O61----WA, and 70.1%, 10.5%, 7.4% and 2.7% for WA----WB). Averaged over all water molecules, the difference between experimental and MD water positions of 0.072 nm is larger than that for the α-cyclodextrin atoms, which was 0.025 nm. A similar observation is found for the overall atomic fluctuations; the water molecules displayed a value of 0.070 nm (experimental 0.047 nm) and the α-cyclodextrin atoms a value of 0.049 nm (experimental 0.036 nm). Because of the higher mobility of water molecules, it is not surprising that their average positions are less well reproduced in the simulation.

One may compare these results with those of previous simulation studies (Mezei *et al.*, 1983; Vovelle *et al.*, 1985), and distinguish three types of agreement: if in all asymmetric units the deviation between experiment and simulation is smaller than 0.06 nm, there is full agreement; if this is only true in at least one asymmetric unit, it is called partial agreement; if none of these cases was true, it is called no agreement. In the deoxycytidylyl-3′,5′-guanosine–proflavin crystal, the following was found: full agreement in 5 positions; partial agreement in 11 positions; no agreement in 11 positions. When applying the same criterion to our α-cyclodextrin simulation, we found full agreement for three sites (W1, W2, W4) and partial agreement for the other three sites. In the simulation study on the vitamin B12 crystal structure the RMS deviation between simulated and experimental water positions ranges from 0.075 nm to 0.126 nm for various force fields (Vovelle *et al.*, 1985). The value of 0.072 nm found here for the α-cyclodextrin hexahydrate crystal was lower. One may expect better agreement in the α-cyclodextrin case, because the α-cyclodextrin crystal contains fewer water sites per asymmetric unit. However, in the α-cyclodextrin MD simulation all the atoms, including those of the cyclodextrin, are allowed to move, whereas in the deoxycytidylyl-3′,5′-guanosine–proflavin and vitamin B12 studies only the water molecules could move. They may have been positionally restrained by the rigidity of the solutes. Therefore, compared with previous studies the reproduction of the water sites in the crystal lattice of α-cyclodextrin hexahydrate was quite satisfactory.

Hydrogen Bonds

The neutron diffraction structure and the averaged MD structure ⟨M1–16; 5–15 ps⟩ have a similar hydrogen bond network. A hydrogen bond is defined by the distance H----A <0.25 nm and angle D—H----A >135° where A and D are the acceptor and donor atoms. The most important result is that the hydrogen bonds, which have a high probability of occurrence in the MD trajectory, have also been located experimentally. Bonds with lower populations, which exist only for short periods, may correspond to a variety of structures as found in molecular dynamics simulations. They are not described in the experimental paper because molecular dynamics simulations lead to a variety of structures. A more detailed description of the hydrogen bonds may be found in the original publication (Koehler *et al.*, 1987b).

Discussion

The neutron diffraction structure of α-cyclodextrin hexahydrate, form I, and the averaged molecular dynamics structure ⟨M1–16; 5–15 ps⟩ for 16 α-cyclodextrin + 96 water molecules simulated for a period of 15 ps at $T = 293$ K are very similar. The α-cyclodextrin atoms have on average only been displaced by about 0.025 nm, four waters by about 0.03 nm, and water W3 and water WA by 0.12 nm. Fluctuations calculated from experimental *B*-values and from the MD simulation agree in most cases.

The average of the root mean-square positional differences between experimental and MD structures is less than or of the order of the experimentally obtained RMS fluctuations (0.03 nm ring atoms, 0.04 nm side chains; Figure 3.3) and for the water molecules it is about 0.07 nm. The main characteristics are well reproduced such as the twisted glucose unit 5 with a glycosidic bond torsion angle of about 158° (experiment 171°) instead of 120° and the hydrogen bond from O65 to the included water WA with an occurrence of 91%. The five intramolecular hydrogen bonds between O2 and O3 hydroxyl groups of adjacent glucose units and the large O25–O36 distance observed experimentally are also found in the MD simulation, i.e. the α-cyclodextrin molecule had the tendency to remain in the 'collapsed' form typical for the water inclusion complex and did not prefer the 'round, relaxed' structure observed in inclusion complexes with guest molecules other than water.

Atoms with multiple sites such as O61A and O61B as obtained from neutron diffraction refinements show large MD fluctuations, which means that the crystallographic disorder phenomenon is explained by a high mobility of the corresponding atom in the MD simulation. In general, hydrogen bonds with occurrence above 50% of the time in the MD study are also observed experimentally. A striking example where all geometrically possible hydrogen bonds between two water molecules are present in the MD simulation is found with the two included waters WA and WB. They interacted also with glucose O4 oxygen atoms at a low percentage (<10%), which cannot be seen experimentally.

Finally, we conclude that the experimental properties of the α-cyclodextrin hexahydrate crystal are well reproduced by the force field that is used in the MD simulation presented here. It could well be that the experimental data obtained with the 'hedgehog', a 100-counter diffractometer, are not good enough to provide a really reliable basis for this MD study because the reliability factor for the refinement is only 11%. Therefore, new neutron diffraction data have to be collected in order to find out whether the hydrogen bonds simulated with lower occupancy really have a physical meaning.

MOLECULAR DYNAMICS SIMULATION OF CRYSTALLINE β-CYCLODEXTRIN DODECAHYDRATE AT 293 K AND 120 K

The seven-membered species in the class of cycloamyloses with glucose

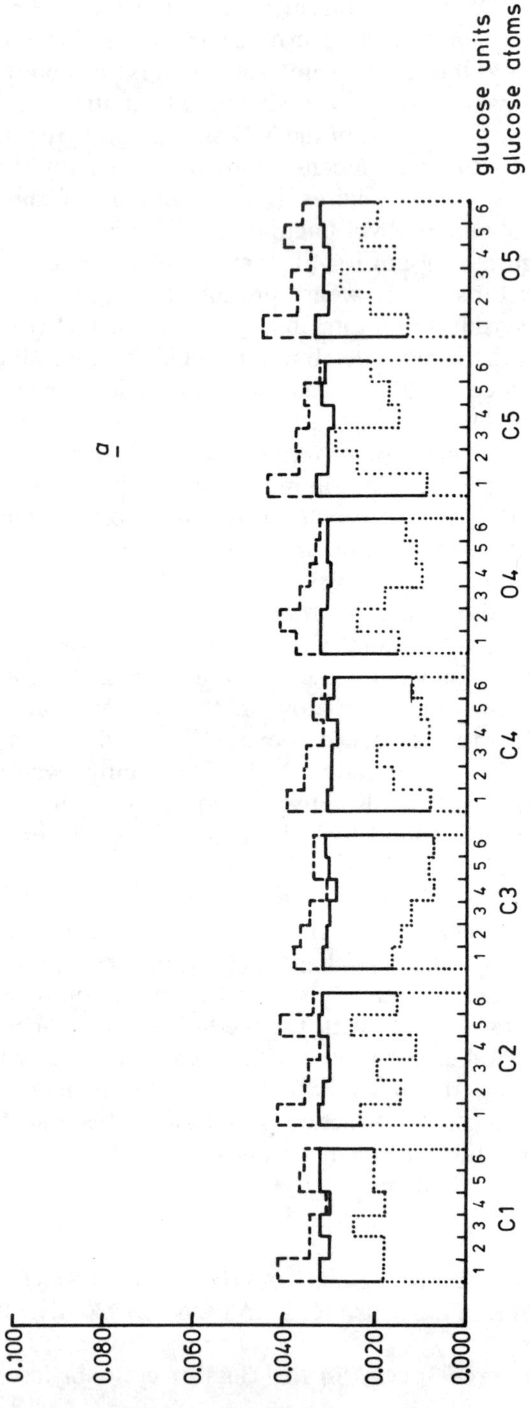
0.100
0.080
0.060
0.040
0.020
0.000
a
1 2 3 4 5 6
C1
C2
C3
C4
O4
C5
O5
glucose units
glucose atoms

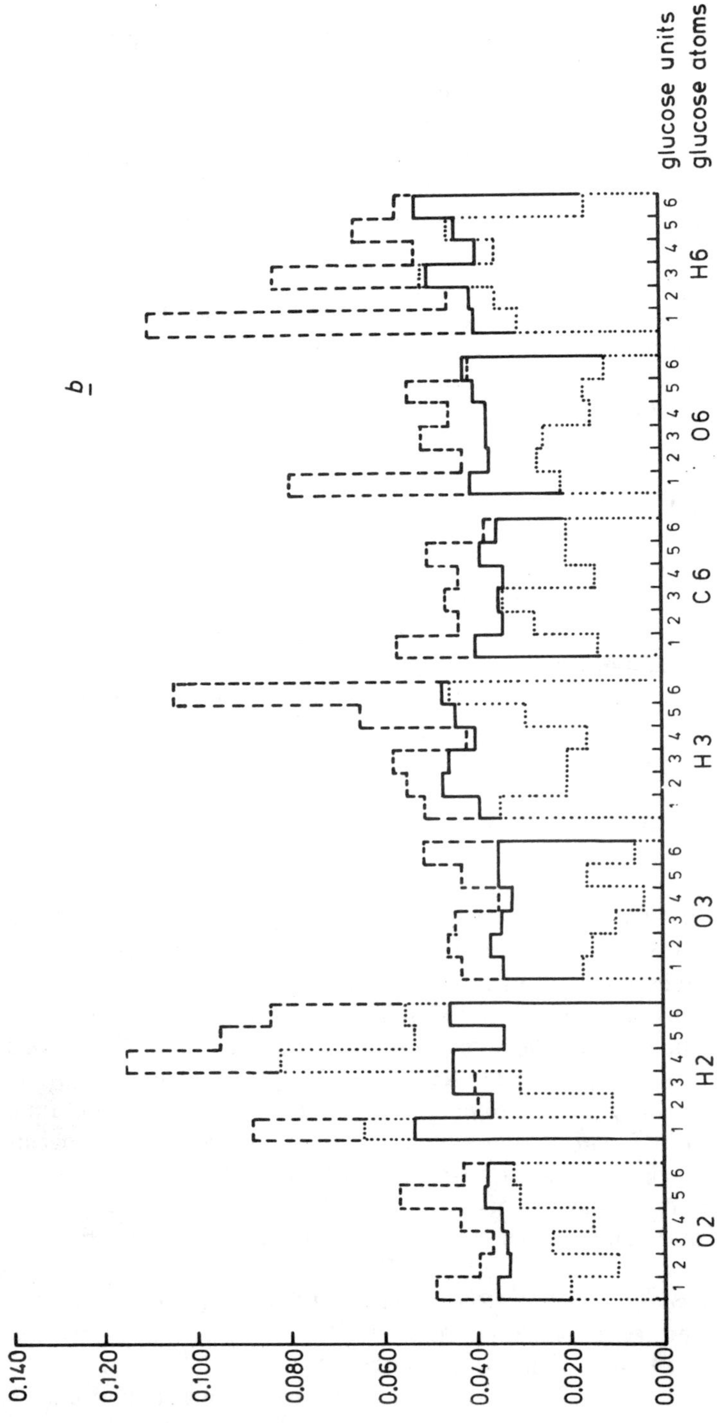

Figure 3.3 RMS positional fluctuations and differences (in nm) of (a) main chain and (b) side chain atoms in the six glucose units. Solid line, fluctuations as derived from experimental isotropic *B*-factors; dashed line, fluctuations obtained from the MD simulation ⟨M1–16; 5–15 ps⟩; dotted line, differences between MD averaged structure and experimental structure

units linked by α-1–4 glycosidic bonds is called β-cyclodextrin. Its central cavity has a diameter of 0.6–0.64 nm and it is able to include organic and inorganic guest molecules (Saenger, 1980; Atwood *et al.*, 1984). The room temperature form contains extended flip–flop hydrogen bond networks among water molecules and hydroxy groups of the sugar molecules (Saenger *et al.*, 1982), most of which do not exist at low temperature.

In this study the main question is how well the water molecules are modelled, because β-cyclodextrin contains more water molecules per glucose unit (most of which are disordered) than α-cyclodextrin hexahydrate. Another very interesting question to be answered from this molecular dynamics simulation is how well the potential function, which is an effective force field developed for application at room temperature, functions at 120 K.

Method and Computational Details

The experimental data referred to in this study are the neutron diffraction structures of β-cyclodextrin·$11H_2O$ at 293 K (Betzel *et al.*, 1984) and of β-cyclodextrin·$11.6H_2O$ at 120 K (Zabel *et al.*, 1986).

At both temperatures the space group is $P2_1$ and the number of molecules per unit cell $Z = 2$. The cell dimensions are comparable:

1. 293 K: $a = 2.1261$ nm, $b = 1.0306$ nm, $c = 1.5123$ nm, $\beta = 112.3°$
2. 120 K: $a = 2.1617$ nm, $b = 1.0026$ nm, $c = 1.4891$ nm, $\beta = 112.52°$

The R-values from the full matrix least-squares refinement are $R = 6.8\%$ (293 K) and $R = 4.9\%$ (120 K).

The 11 ± 0.5 crystal water molecules are found to be distributed over 16 oxygen and 37 hydrogen positions at room temperature and over 12 oxygen and 26 hydrogen positions at 120 K, implying that some sites have occupancy factors less than 1.

For the molecular dynamics simulations (Koehler *et al.*, 1987c), 'computational boxes' had been chosen with the following dimensions: (1) 293 K: a, $2b$, $2c$, β; (2) 120 K: a, $2b$, $2c$, β. For each β-cyclodextrin molecule 12 water molecules have been selected in both the room and low temperature MD simulation studies in order to model the experimental sites. This is close to the 11.6 water molecules found in the 120 K neutron diffraction study and to the 11.0 water molecules in the 293 K analysis where, as outlined in Zabel *et al.* (1986), some of the more spurious water sites were probably not seen in the Fermi density maps. The initial water sites are selected according to occupancy and hydrogen bonding. Application of symmetry transformations led to boxes containing 8 β-cyclodextrin molecules (784 atoms) and 8×12 water molecules (288 atoms).

The molecular dynamics simulation is carried out in periods ranging from 0 to 19 ps for 293 K and from 0 to 20 ps for 120 K. The same methodological procedure as described for α-cyclodextrin is used. In order to compare the molecular dynamics structures of β-cyclodextrin with the experimental one, they are averaged over the last 15 ps of the MD run and over all eight molecules (M), referred to as ⟨M1–8; 4–19 ps⟩ for 293 K, and ⟨M1–8; 5–20 ps⟩ for 120 K.

Results

At room temperature, the experimental positions of the (non-hydrogen) glucose atoms were reproduced within 0.034 nm. This value is smaller than the root mean-square atomic fluctuation of 0.041 nm as derived from the crystallographic *B*-factors. The simulation showed an overall RMS atomic fluctuation of 0.049 nm, slightly larger than the experimental value. At low temperature the experimental positions were reproduced within 0.046 nm. This value is larger than the RMS atomic fluctuation of 0.019 nm (experiment) or 0.022 nm (MD). Both experiment and simulation showed a reduction by a factor 2 of the atomic mobility when lowering the temperature from 293 K to 120 K. The larger deviation of the atomic positions from the experimental ones at low temperature might be due to the fact that the empirical interatomic potential function has been designed as an effective force field at room temperature; see Figure 3.4 as an example.

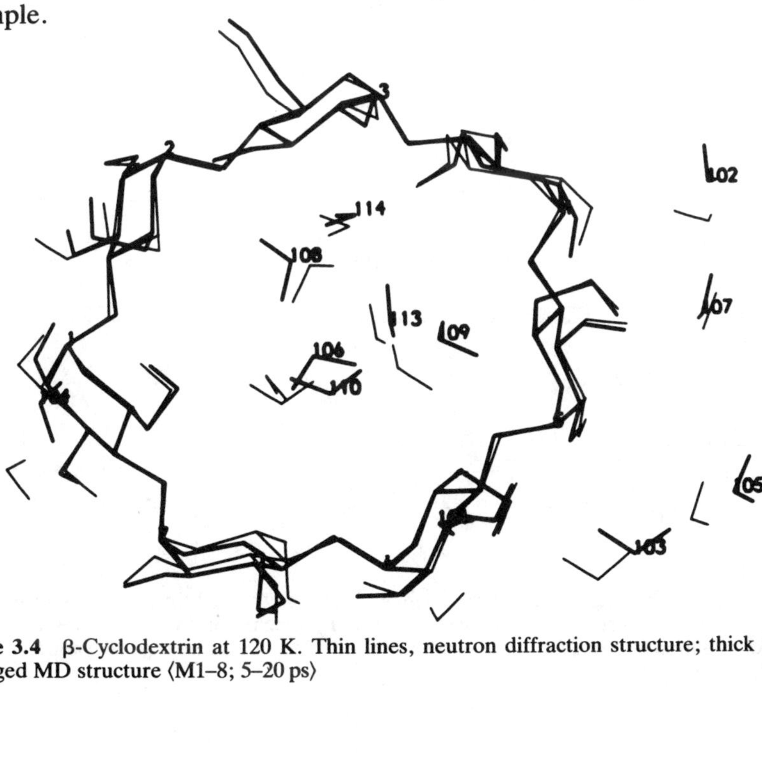

Figure 3.4 β-Cyclodextrin at 120 K. Thin lines, neutron diffraction structure; thick lines, averaged MD structure ⟨M1–8; 5–20 ps⟩

In the average values of bond angles and torsion angles, the slight changes that are experimentally observed when reducing the temperature were qualitatively reproduced by the simulation. Also, the variation of corresponding torsion angles along the cyclic chain of 7 glucose units in β-cyclodextrin was qualitatively mirrored in the simulations. At both temperatures the degree of anisotropy in the atomic motions that is observed experimentally was also found in the simulations. At room temperature, the neutron diffraction refinement yields alternative sites for 17 glucose atoms, one O6 oxygen and 16 hydroxyl hydrogens. For

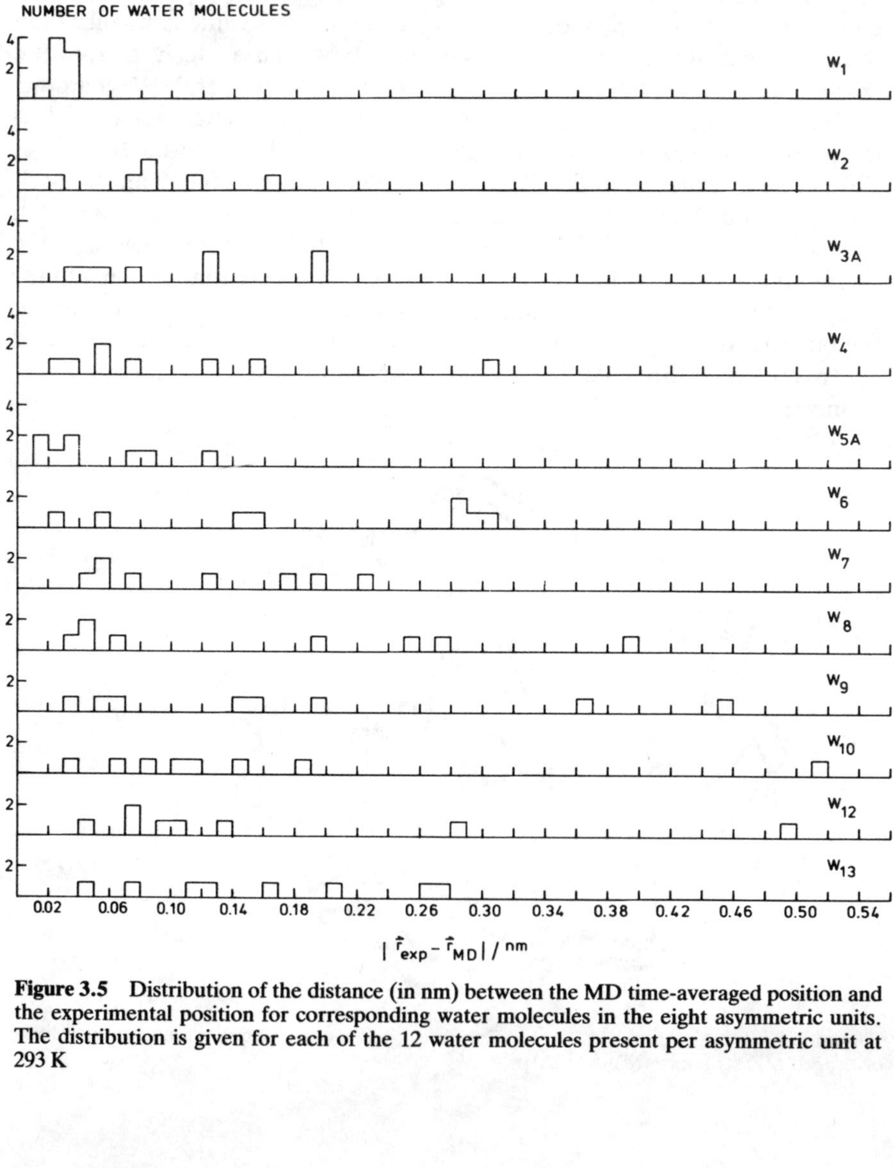

Figure 3.5 Distribution of the distance (in nm) between the MD time-averaged position and the experimental position for corresponding water molecules in the eight asymmetric units. The distribution is given for each of the 12 water molecules present per asymmetric unit at 293 K

two-thirds of the atoms the relative occupancy of the alternative sites was qualitatively reproduced in the simulation. Although the β-cyclodextrin crystal contains twice as many water molecules per cyclodextrin molecule compared with the α-cyclodextrin crystal and also showed more atomic

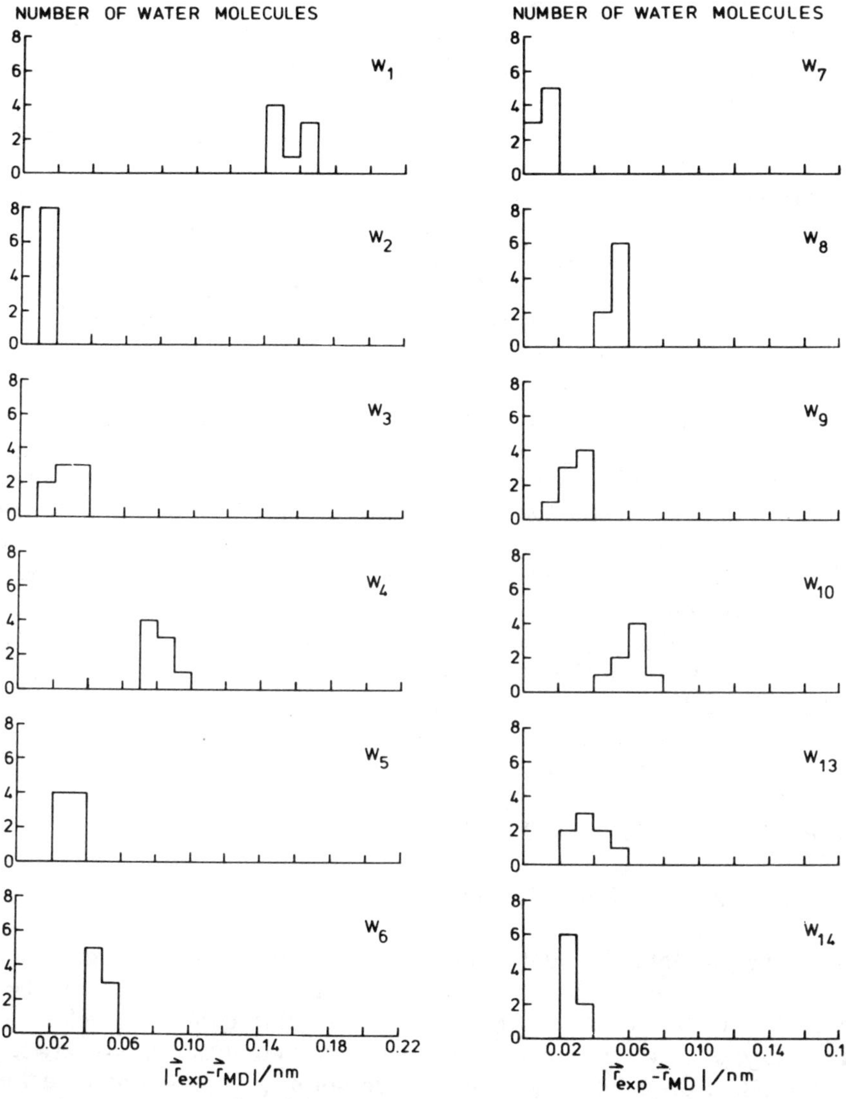

Figure 3.6 Distribution of the distance (in nm) between the MD time-averaged position and the experimental position for corresponding water molecules in the eight asymmetric units. The distribution is given for each of the 12 water molecules present per asymmetric unit at 120 K

disorder, the agreement with the experimental data for the glucose atoms was of the same quality in both studies.

From Figures 3.5 and 3.6 it can be seen how far the water molecules could move in the simulation at room temperature and how much they stayed close to their original position at low temperature. At low temperature, for 10 out of 12 water molecules the simulated positions lay within 0.06 nm from the crystallographic positions. Only water W1 is consistently shifted away from its experimental position by 0.15 nm in all eight asymmetric units. The simulated RMS atomic fluctuations of the water molecules (0.03 nm) are slightly smaller than the experimental ones (0.04 nm). The water molecules in the different asymmetric units show very similar behaviour.

At room temperature, the crystallographic refinement yields 16 oxygen and 37 hydrogen sites which are occupied by 11 ± 0.5 water molecules. Only three sites (W1, W2, W4) are fully occupied. Therefore, it is not surprising that the simulation displays a larger overall RMS positional fluctuation for the water atoms of 0.10 nm. The value derived from the crystallographic *B*-factors is 0.053 nm. The discrepancy between these values is explained by the occurrence of alternative sites which are accessible for one atom. The crystallographic refinement treats different sites that are partly occupied by one atom separately with a *B*-factor for each site, whereas the MD values are time averaged over the trajectories of individual atoms, irrespective of the different sites that are visited by one atom. Only when a water molecule does not diffuse from one site to another in the simulation can its time-averaged position be compared with the experimental one. Only water W1 has a simulated mobility (0.040 nm) that is comparable with the experimental value (0.044 nm). The experimental position of this water molecule is reproduced within 0.03 nm. For all other water molecules their diffusion does not allow a direct comparison of time-averaged and experimental positions.

FLIP–FLOP HYDROGEN BONDING

Method and Computational Details

In the analysis of the MD data, our previous definition of a hydrogen bond is used (i.e. H---A <0.25 nm and angle D—H---A >135°). When applied to each time frame of the MD trajectories this definition yields the percentage of occurrence of a hydrogen bond such that transitions between different hydrogen bond types can be monitored. Transitions of a hydrogen bond from one to another configuration sometimes show a diffuse pattern: atomic positional fluctuations at the transition state may generate many counted transitions owing to a strict application of the hydrogen bond criterion. Therefore we have filtered out the effect of the local fluctuations by counting a transition only when a specific hydrogen bond has ceased to exist during 10 time frames (0.2 ps). In this way hydrogen bond lifetimes are calculated which have a lower bound of 0.2 ps and an upper bound of 10 ps (α-cyclodextrin) or 15 ps (β-cyclodextrin), i.e. the time period covered by the MD simulations. It should be mentioned here again that the integration time step during the simulation is 0.002 ps, and every 0.02 ps a configuration time frame is stored for the analysis of hydrogen bonds as function of time (Koehler *et al.*, 1988a).

Hydrogen Bonds in β-Cyclodextrin Hydrate at 293 K and 120 K

In the MD simulations, the water molecules can move quite freely in some locations, whereas in others they stay in approximately the same area. At

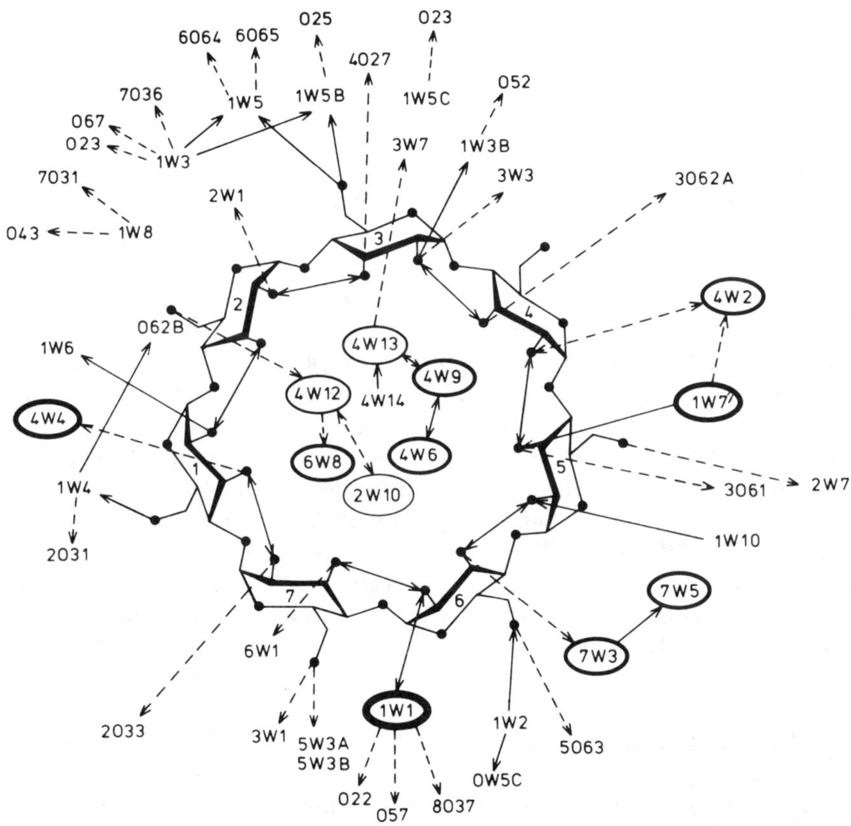

Figure 3.7 Hydrogen bonds of β-cyclodextrin at 293 K as obtained from neutron diffraction data

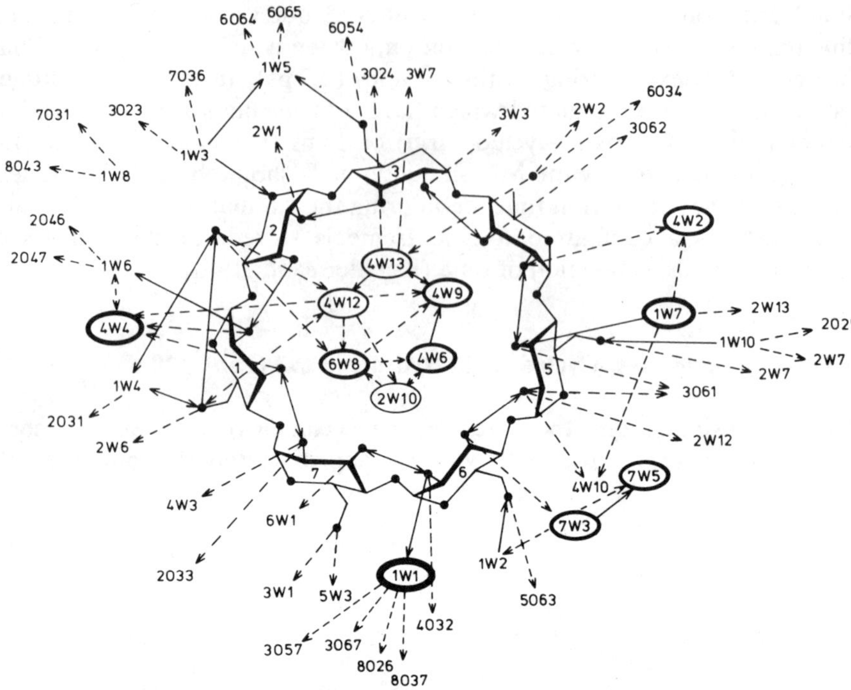

Figure 3.8 Hydrogen bonds of β-cyclodextrin at 293 K from the MD time-averaged ⟨4–19 ps⟩ data

293 K, from the 96 water molecules in the computational box 7 remain close to the experimental site. Nineteen waters drift away and move into each other's original sites. Therefore, the simulation predicts many hydrogen bonds with low occurrence in the 1–30% range, which are found between the glucose and water atoms and among the water molecules. These hydrogen bonds, even if they really occur, cannot be seen in the neutron diffraction experiments. This is because hydrogen atoms with such low occupancy are indistinguishable from background noise in the nuclear density map. Of course, the strong hydrogen bonds, in most cases, are found in the experimental structure as well as in the simulated one, as can be seen in Figures 3.7 and 3.8 for the 293 K study and in Figures 3.9 and 3.10 for the 120 K study.

Flip–Flop Hydrogen Bonding

In β-cyclodextrin at 293 K there is an infinite flip–flop chain between O37(1) ↔ O26(1) ↔ OW1(1) ↔ O37(8) etc., which is formed through operation of the 2_1 screw axis. The MD calculation shows that in the

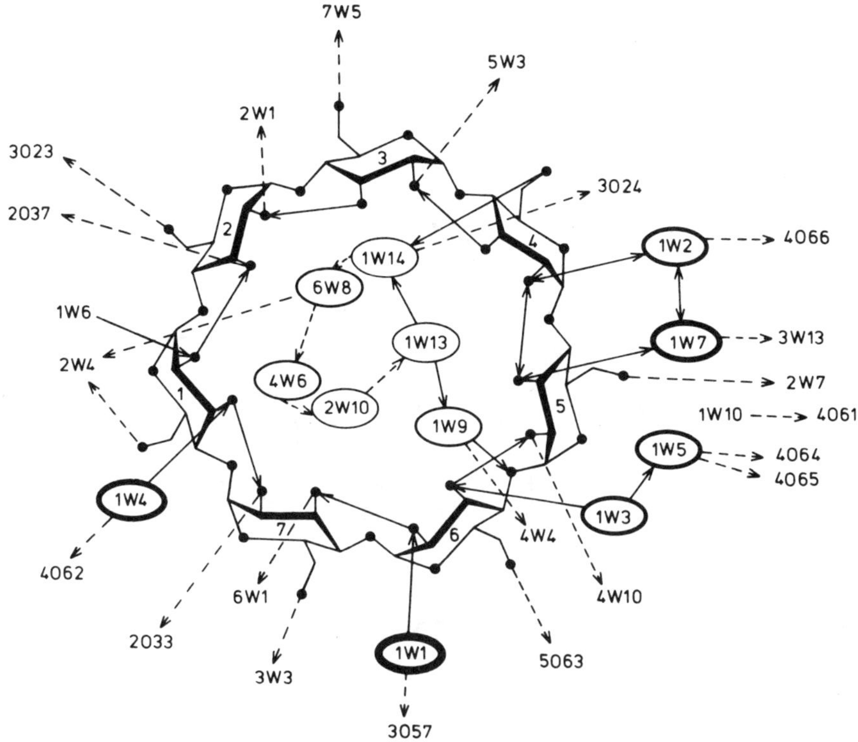

Figure 3.9 Hydrogen bonds of β-cyclodextrin at 120 K as obtained from neutron diffraction data

simulated eight β-cyclodextrin asymmetric units (four unit cells), there are two infinite chains formed by these atoms. The first chain runs through asymmetric units 1, 8, 3, 6 and the second through 2, 5, 4, 7. During the 15 ps analysis period of the simulations both chains are independent and display different orientations of hydrogen bonds. A superposition of the data of this system consisting of eight groups with three hydroxyls each leads to an average that shows flip–flop characteristics. This example illustrates that the molecules in the eight different asymmetric units behaved independently during the simulation.

These results demonstrate that the GROMOS force field is able to reproduce satisfactorily the experimentally observed flip–flop hydrogen bonds in β-cyclodextrin at 293 K. At low temperature (120 K), the four-membered flip–flop cycle remains in its starting configuration, which is reproduced in MD with three of these bonds, whereas the fourth flip–flop occurs only with a very low percentage (1–2%), because water 2 has moved slightly in the simulation. This four-membered cycle is also analysed in the β-cyclodextrin structure at 293 K.

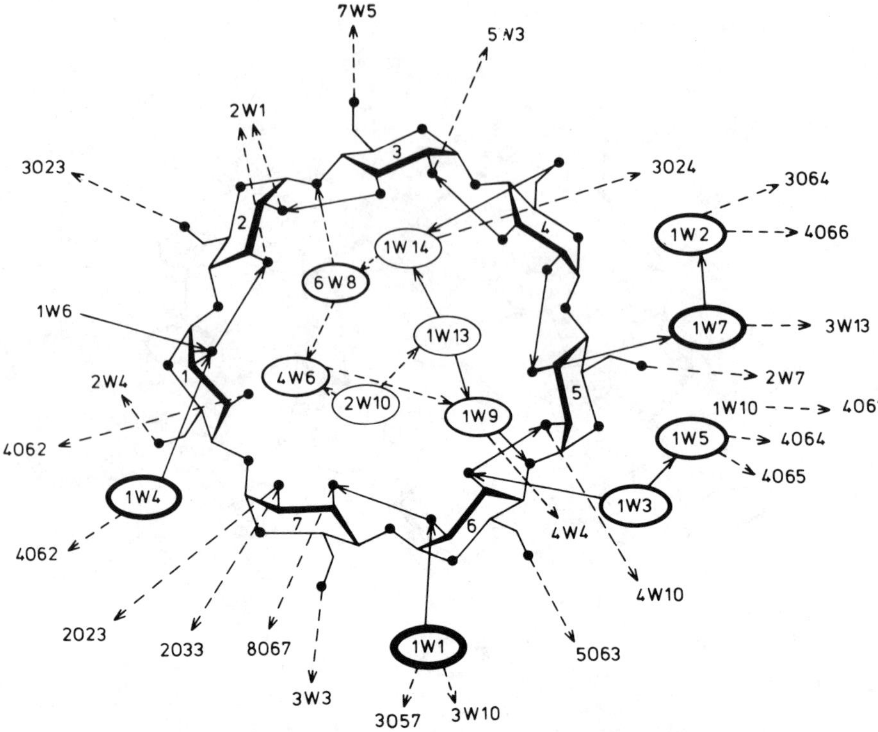

Figure 3.10 Hydrogen bonds of β-cyclodextrin at 120 K from the MD time-averaged ⟨5–20 ps⟩ data

The main characteristic of the simultaneous occurrence of four hydrogen bonds in the four-membered cycle O24(1)–O35(1)–OW7(1)–OW2(4)–O24(1) in β-cyclodextrin at 293 K (Figure 3.11) is that they have short lifetimes in the MD simulation. They change to a different configuration by changing only one bond at the time, and not as expected by a concerted cooperative flip. During the 15 ps simulation there are eight time periods (longer than 0.05 ps) where the four-membered cycle persists in β-cyclodextrin molecule 1. The lengths of these time periods are all shorter than 0.08 ps with an average length of 0.036 ps. Molecules 2 and 4 do not show the four-membered cycle during the whole 15 ps. In molecule 5 the sum of all 17 time periods where hydrogen bonding occurred is 4.14 ps; this illustrates how large the differences between the trajectories of the molecules can be. The average lifetime of the four-membered cycle is about 0.04 ps. The directions of the hydrogen bonds in the cycle change during the simulation. In the simulation, the four-membered cycle exists only for 5.4% of the time. This low percentage would not allow one to detect the flip–flop cycle experimentally at 293 K. Nevertheless, at 293 K,

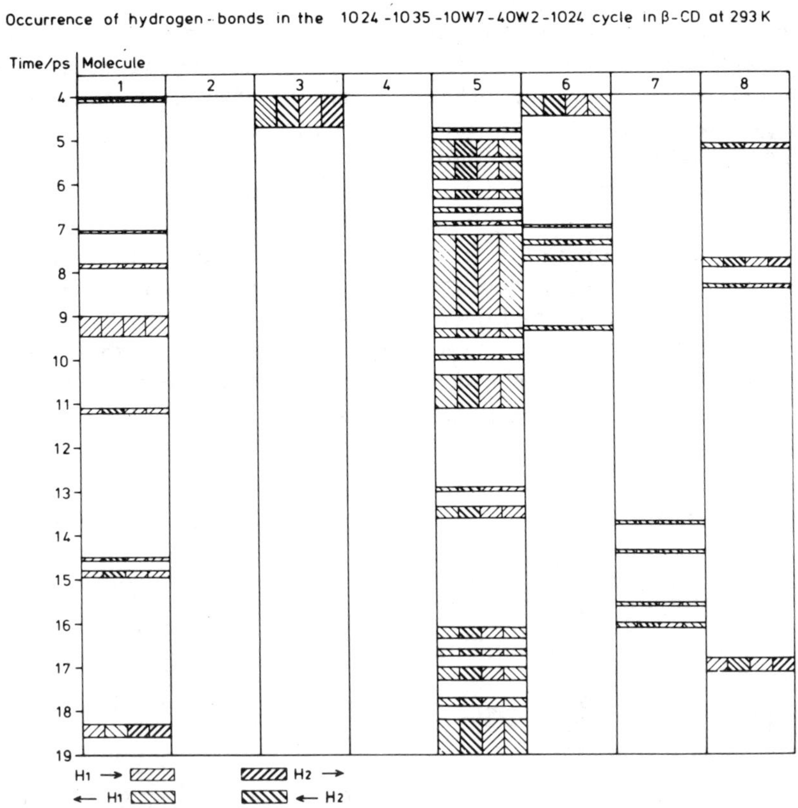

Figure 3.11 Hydrogen bonds of the O24(1)–O35(1)–OW7(1)–OW2(4)–O24(1) cycle in β-cyclodextrin at 293 K. Occurrence (larger than 0.05 ps) of hydrogen bonds during the MD simulation

all four bonds in this cycle are of the flip–flop type in the MD study, whereas in the experiment there are two bonds in the cycle of normal O—H----O type.

At 120 K, this cycle changes to the flip–flop type O24(1) ↔ O35(1) ↔ OW7(1) ↔ OW2(1) ↔ OW24(1) which was well established experimentally. In the MD simulation, however, it exists only during very short time intervals. The orientations of its hydrogen bonds are always H1→H1→H2→H2→. The overall occurrence in all eight asymmetric units is only 0.32 ps during the simulated 15 ps (= 0.04 ps per molecule per 15 ps). Most of the time there were only three hydrogen bonds: H1→H1→H2→O. The MD simulation in this case does not reproduce the experiment to our satisfaction. In summary, at 120 K, the neutron diffraction structure of β-cyclodextrin · $11H_2O$ displays only four flip–flop hydrogen bonds which are arranged in a cycle. The MD simulates only three of these bonds, but in a unidirectional mode, and the fourth exists only occasionally and then closes the cycle as a normal O—H----O hydrogen bond. This is not too surprising considering the low temperature and the time period of only 20 ps used for the simulation.

The experimentally determined infinite flip–flop chain of OW6 ↔ OW9 ↔ OW13 ↔ OW6 in β-cyclodextrin at 293 K is well reproduced in the MD simulation (Figure 3.12).

A comparable analysis is carried out for the two enclosed water molecules A and B in α-cyclodextrin · $6H_2O$. The neutron diffraction study shows a clearly defined hydrogen bond OWA—H----OWB. This bond is very well reproduced in the MD simulation for about 80% of the time, and for only 10% of the time shows a flip–flop with a hydrogen bond in the OWA----H—OWB direction (Koehler *et al.*, 1987b). This contrasts strik-

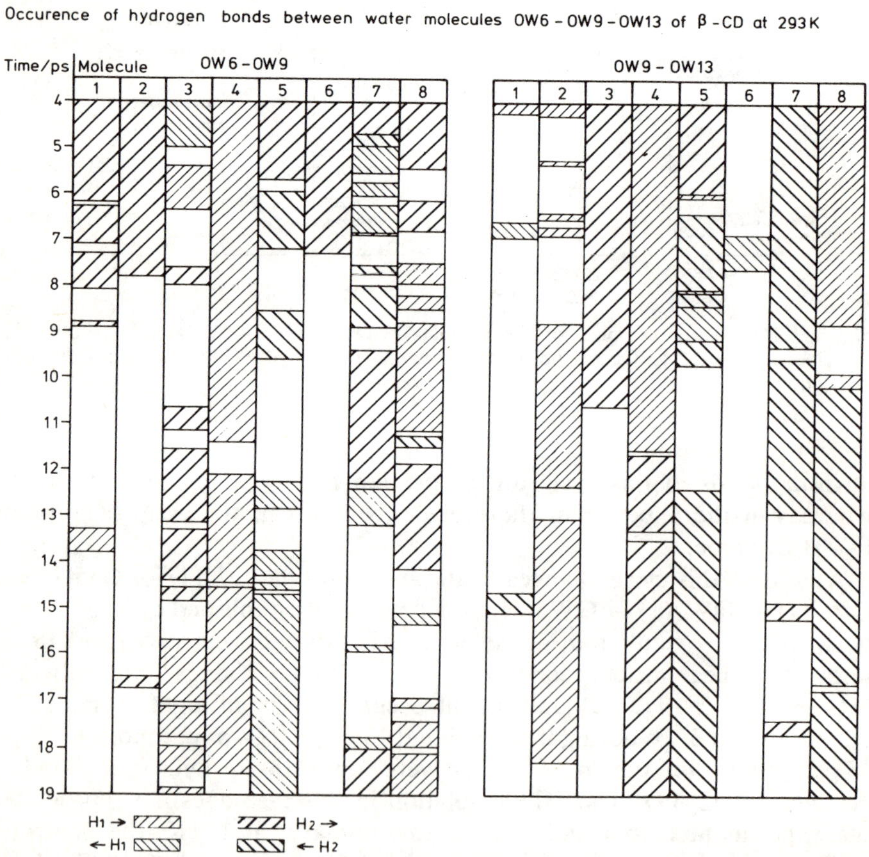

Figure 3.12 Hydrogen bonds between three water molecules OW6(1)–OW9(1)–OW13(1) of β-cyclodextrin at 293 K

ingly with the β-cyclodextrin hydrate system, where the wider cavity allows the enclosed water molecules to be disordered and to show extensive flip–flop hydrogen bonds.

In general, the GROMOS force field yields a good description of the cyclodextrin hydrate crystal structures at room temperature. The dynamic behaviour of the flip–flop hydrogen bonds is well reproduced qualitatively (84%), and often quantitatively. At 120 K, some discrepancies occur with the flip–flop dynamics but the normal, static O—H----O hydrogen bonds calculated by MD agree with the experimental observations.

CONFORMATIONAL DIFFERENCES BETWEEN α-CYCLODEXTRIN IN AQUEOUS SOLUTION AND IN CRYSTALLINE FORM

The computer simulation techniques of molecular dynamics are a powerful tool to delineate the conformational differences between a molecule in different environments. As an illustration, the difference between an α-cyclodextrin molecule in aqueous solution and in crystalline form is determined. Here we analyse the behaviour of α-cyclodextrin simulated in an aqueous environment (Koehler *et al.*, 1988b).

Method and Computational Details

α-Cyclodextrin in aqueous solution is modelled by simulating one α-cyclodextrin molecule in a truncated octahedron containing 611 water molecules. The coordinates of one α-cyclodextrin and six water molecules were taken from the asymmetric unit of the form I crystal (Klar *et al.*, 1980) and put in the centre of the truncated octahedron. The additional 605 water molecules are inserted into the computational box by immersing it in an equilibrium configuration of bulk SPC water and subsequently removing all water molecules that are outside the truncated octahedron. Any close non-bonded contacts are relaxed by performing energy minimization in which all the α-cyclodextrin atoms are restrained to their initial positions. No restraints are applied during MD simulation. In order to allow for a relaxation of the density of the system, the MD simulation is performed at constant pressure of 1 atm (=101 325 Pa). The pressure relaxation time was $\tau_P = 0.5$ ps and the compressibility of 0.7624×10^{-3} (kJ mol^{-1} $nm^{-3})^{-1}$ is that of water at $T = 293$ K. The temperature is kept constant as in the crystal simulation by a weak coupling ($\tau_T = 0.1$ ps) to a temperature bath of $T_0 = 293$ K. Newton's equations of motion are integrated with a time step $\Delta t = 2$ fs over a period of 90 ps. The potential energies stabilize within a few picoseconds, so the final 80 ps of the trajectory can be used for analysis. This yields statistics about half as good for α-cyclodextrin in solution as for α-cyclodextrin in crystalline form, where 10 ps trajectories of 16 molecules were used. The crystal simulation is performed at constant volume whereas the solution simulation is carried out at constant pressure.

Results

The solution structure of α-cyclodextrin deviated by about 0.1 nm from that in the crystal and showed twice as much mobility of the atoms. The experimentally observed twist of glucose unit 5 out of alignment with the other five glucose units in the α-cyclodextrin torus that was present in the crystal simulation disappeared in the simulation in solution (see Figure 3.13) but the glycosidic torsion angles around the ring remained asymmetric.

In the simulation the intramolecular hydrogen bonds are less stable in solution than in the crystal, as judged from their populations. In solution some of the intramolecular hydrogen bonds show so-called flip–flop character. The hydrogen bond lengths H----A seem to be slightly shorter in solution than in the crystal. The hydrogen bonds between α-cyclodextrin and water showed rather different patterns in solution and in the crystal. Hydrogen bonds that are present in the crystal are observed in solution, but a number of hydrogen bonds in solution are not observed in the crystal. The crystal shows only one static hydrogen bonding pattern out of a variety of patterns that can be formed in aqueous solution. This study shows that the crystal structure is only one member of a wide family of structures that are accessible to α-cyclodextrin in solution. In aqueous solution the molecule clearly visits a larger part of conformation space than it did in the crystal.

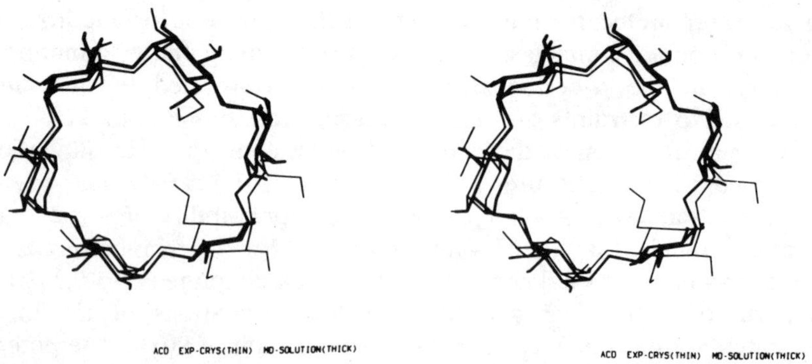

Figure 3.13 α-Cyclodextrin at 293 K; crystal structure (thin lines) and MD solution structure (thick lines) (stereo diagram)

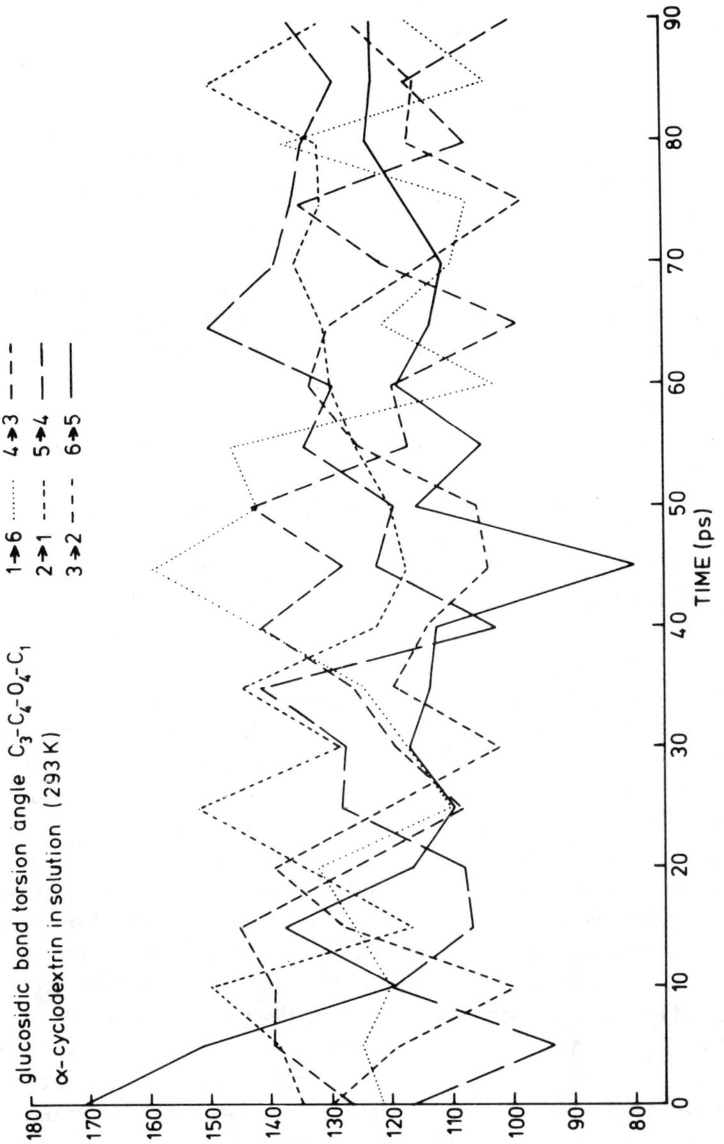

Figure 3.14 Time evolution from 0 to 90 ps of the six glycosidic bond torsion angles C3–C4–O4–C1 of α-cyclodextrin in solution

The glycosidic bond torsion angles show interesting conformational differences between solution and crystalline forms. Experimentally, the values lay in the range 115–135° with one exception, the torsion angle connecting glucose units 5 and 6, which took the value of 171°. In the crystal simulation this twist in glucose unit 5 is reproduced. In solution, the twist disappears within a few picoseconds and the six glycosidic bond torsion angles show comparable behaviour. Their average value lay in the range 115–135°, with fluctuations considerably larger in solution than in the crystal. Incidentally, from Figure 3.14 it can be observed that the glycosidic torsion angles show large excursions (±40°) from their mean value, which indicates that the six-membered α-cyclodextrin macrocycle is intrinsically asymmetric as a result of internal strain, and that the occurrence of one twisted glucose unit in the crystal is energetically not unfavourable. In terms of the glycosidic bond torsion angles α-cyclodextrin adopts a twofold symmetry in solution; glucose units 2 and 5 show lower values (116°) than the others.

While the distribution of individual glycosidic torsion angles in the α-cyclodextrin molecule describes the rotational motion of the glucose units with progress of time, it provides no direct information on the intramolecular interglucose O(2)----O(3) hydrogen bonding, which is of major importance for the conformational stability of the cyclodextrins. For an α-cyclodextrin molecule with all six O(2)----O(3) hydrogen bonds at the ideal 0.285 nm distance, we would expect the sum of the six hydrogen bonds to be 1.71 nm. The time course of the six O(2)----O(3) hydrogen bonds for α-cyclodextrin in solution shows individual fluctuations from 0.263 nm to 0.400 nm if we omit the crystallographically observed extremes of 0.251 nm and 0.439 nm at the start of the simulation, but the sum of the six distances is in a narrow range, 1.805–2.038 nm, or the mean O(2)----O(3) distance is in the range 0.30–0.34 nm. The α-cyclodextrin molecule has the possibility of adopting a regular structure with all O(2)----O(3) distances at 0.30–0.35 nm, corresponding to rather weak hydrogen bonds. This is obviously unfavourable and at all times one or two of the hydrogen bonds are broken (O(2)----O(3) distances >0.35 nm) and the others are formed at their expense at shorter than average distance. This is the most preferred situation and is reflected in the two α-cyclodextrin·$6H_2O$ crystal structures with 'collapsed' macrocycle conformation (see p. 74). At certain instances (at 15 ps, 27–30 ps, 50 ps, 73 ps and 80 ps) all the O(2)----O(3) distances are below 0.35 nm and we can assume that the ring of hydrogen bonds is really formed. This situation appears to represent a transition intermediate in solution, as it occurs less frequently than the 'collapsed' form. It had been observed crystallographically in α-cyclodextrin·$7.57H_2O$ where, in contrast to the hexahydrate complex, the water in the cavity is disordered and the α-cyclodextrin adopts the 'round' structure observed also in complexes with guest molecules. All six hydrogen bonds are well defined, since the hydrogen atoms are located, and the O(2)----O(3) distances are in the glucose order 1–2, 2–3, etc.: 0.2941 nm, 0.2900 nm, 0.2913 nm, 0.2974 nm, 0.3150 nm, 0.3011 nm, with the sum of distances at 1.7889 nm and the mean O(2)----O(3) distance at 0.2982 nm (Chacko and Saenger, 1981). For the β- and γ-cyclodextrins, the curvature of the macrocycle is smaller than that of α-cyclodextrin, and the average O(2)----O(3) distances are also smaller: 0.286 nm and 0.281 nm (Lindner and Saenger, 1980). Therefore, strong and well-defined interglucose O(2)----O(3) hydrogen bonds can form, conferring extra stability and rigidity to the macrocycle. With these larger cyclodextrins, molecular dynamics simulations are therefore expected to exhibit only modest excursions with time from the average, lowest energy minimum structure.

Finally, we note that the properties of the six glucose units in α-cyclodextrins in solution should be identical, because of symmetry considerations. For example, the six averaged torsion angles O(5)–C(1)–C(2)–O(2) and fluctuations are about equal; the C(1)–C(2)–O(2)–H(2) angles show much larger differences, which means that the motion of this torsion angle is not completely covered by the simulation. A similar observation could be made for the hydrogen bonding properties of equivalent atoms in the six glucose units.

THREE-CENTRE HYDROGEN BONDS IN CYCLODEXTRINS

Three-centre (bifurcated) hydrogen bonds may play a role by serving as an intermediate state between different dynamically changing hydrogen bonding patterns. Hydrogen bonding configurations can be studied experimentally by neutron diffraction and theoretically by computer simulation techniques. Here, both methods are used to analyse the occurrence of three-centre hydrogen bonds in crystals of cyclodextrins.

Almost all experimentally observed three-centre hydrogen bonds in the crystal are reproduced in the molecular dynamics simulations (Koehler *et al.*, 1988c), even as far as the detailed asymmetric geometry is concerned. On the basis of this result an MD simulation of cyclodextrin in aqueous solution is searched for the occurrence of three-centre hydrogen bonds. Significant differences are found. In solution more different three-centre hydrogen bonds per α-cyclodextrin molecule are observed than in the crystal but the population (existence as a percentage of the simulation period) of each three-centre hydrogen bond is lower in solution than in crystal. Three-centre hydrogen bonds might indeed serve as intermediate states in the process of changing one hydrogen bonding pattern into another.

A normal two-centre hydrogen bond D—H---A with donor D and acceptor A can be defined in energetic or in geometric terms. In dynamic systems a hydrogen bond donor D may exchange its current acceptor A1 for another acceptor A2 following the pathway sketched in Scheme 3.2, where the configuration of the four atoms is assumed to remain planar.

The hydrogen bond configurations with two acceptors A1 and A2 are called bifurcated or three-centre hydrogen bonds. As a criterion for their occurrence the following conditions must be met: both distances H----A1 and H----A2 are shorter than 0.27 nm, both angles D—H---A1 and D—H----A2 are larger than 90°, the sum of the three angles D—H---A1, D—H----A2 and A1---H----A2 is larger than 340° and the dihedral angle defined by the plane through atoms A1, A2 and H and the plane through A1, A2 and D is smaller than 15°. In rare cases the two parts of the three-centre hydrogen bond are about equal as are the D—H---A angles. Generally they differ, the major component being associated with a shorter H---A distance and a more linear D—H---A angle (denoted by ----), the minor with a longer H---A distance and a more acute D—H-----A angle (denoted by ------).

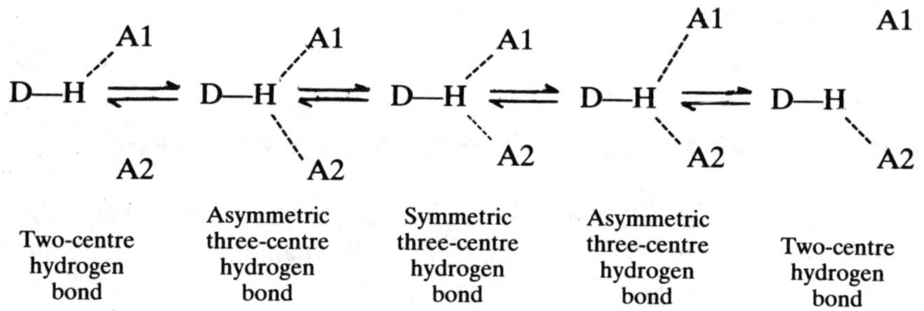

Scheme 3.2

If three-centre hydrogen bonds only occurred as intermediate transition states between stable, normal two-centre hydrogen bonds, they are difficult to detect experimentally. However, in crystals of small biological molecules, such as saccharides, nucleotides and amino acids, they occur statically and can be studied by X-ray and neutron diffraction methods (Ceccarelli *et al.*, 1981; Jeffrey and Mitra, 1984; Jeffrey and Maluszynska, 1986).

By comparing the MD simulation of α-cyclodextrin in crystalline form with a simulation in aqueous solution, it is possible to check whether the three-centre hydrogen bonds occurring in the crystal are maintained when the α-cyclodextrin molecule is 'dissolved' in water.

Method and Computational Details

Here, the trajectories of four different MD simulations which have been described previously are used for analysing hydrogen bonding patterns:

1. the simulation of β-cyclodextrin in a crystal at a temperature of 120 K;
2. the simulation of the same system over a period of 19 ps at 293 K;
3. the simulation of α-cyclodextrin in a crystal at 293 K;
4. a simulation of one α-cyclodextrin molecule in an aqueous solution over a period of 90 ps at $T = 293$ K.

Results

Our data show that the MD simulations of crystalline α- and β-cyclodextrin do reproduce the experimentally observed three-centre hydrogen bonds. At low temperature ($T = 120$ K) only 2 out of 13 experimentally observed three-centre hydrogen bonds are not found in the simulations. A more detailed comparison shows that even the asymmetry in the experimentally observed three-centre hydrogen bonds (Scheme 3.2), characterized by the relations distance(H----A1) < distance(H------A2) and angle(D—H----A1) > angle(D—H------A2), is well reproduced by the simulations. This indicates that the interatomic force field and the simulation techniques that are used are sufficiently reliable to study hydrogen bonding behaviour in the crystalline state and in solution.

The average ratio between the population of simulated three-centre and two-centre hydrogen bonds ranges from 50% to 70% for intramolecular three-centre hydrogen bonds. It is highest for crystalline β-cyclodextrin at 120 K followed by α-cyclodextrin at room temperature in solution and in the crystal, and it is lowest for crystalline β-cyclodextrin at 293 K. In this study we have selected from all crystal structures only the experimentally observed three-centre hydrogen bonds and compared them with the simulated analogues. In the molecular dynamics trajectories in the crystalline state there are many other examples of three-centre hydrogen bonds, which gives the static picture emerging from the experimental study a more dynamic character.

When comparing the simulation of α-cyclodextrin in aqueous solution with the simulation of this molecule in crystalline form (Figure 3.15 is an example; for details see the tables in Koehler *et al.*, 1988c) a number of differences become obvious. In solution many more three-centre hydrogen bonds are observed but they occur only for a short time. They also seemed to be more symmetric than the ones observed in the crystal. The crystal forces seem to stabilize specific three-centre hydrogen bond patterns. The ratio between three-centre and two-centre hydrogen bonds is about 10%. This should be borne in mind when interpreting and generalizing observa-

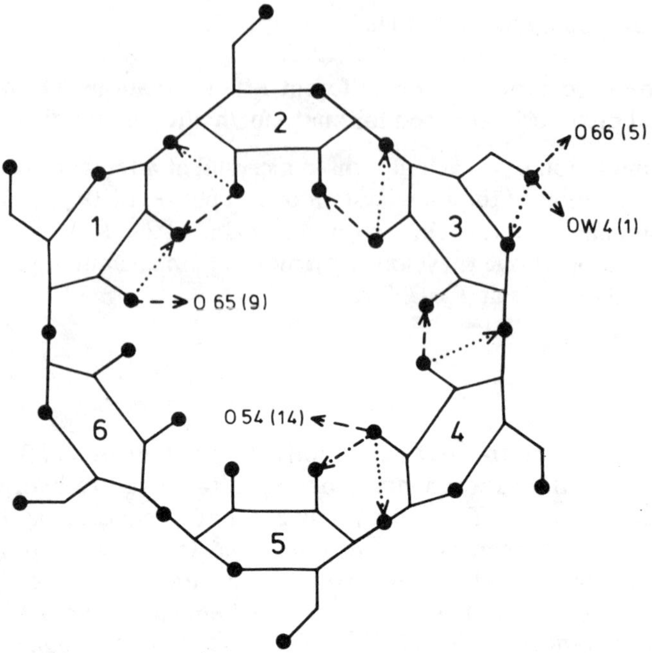

Figure 3.15 Schematic drawing of the crystal structure of α-cyclodextrin (form I) with its three-centre hydrogen bonds at 293 K

tions about hydrogen bonding in crystals to other molecular environments.

Finally, we note that the MD simulations draw a rather dynamic picture of hydrogen bonding patterns of molecules in an aqueous environment. The data show that three-centre hydrogen bonds form intermediate states between different hydrogen bonding patterns and are certainly compatible with the results of this cyclodextrin study. In the hydration shell, water molecules may slide over the surface of a solute by continuously exchanging hydrogen bond donors or acceptors through three-centre hydrogen-bonded intermediate states. Conversely, the dynamical transitions postulated for the flickering cluster model proposed for liquid water can be envisaged to occur smoothly via three-centre hydrogen bond intermediates.

REFERENCES

Ahlstrom, P., Wallqvist, A., Engstrom, S. and Jonsson, B. (1989). A molecular dynamics study of polarisable water. *Mol. Phys.*, **68**, 563–581

Allen, M. P. and Tildesley, D. J. (1987). *Computer Simulation of Liquids*. Oxford Science Publications, Clarendon, Oxford

Allinger, N. L. and Graham, J. C. (1973). Organic quantum chemistry. XXVIII. The bond order–bond length relationship within the VESCF method. *J. Am. Chem. Soc.*, **95**, 2523–2526

Allinger, N. L. and Sprague, J. T. (1972). Conformational analysis. LXXXIV. A study of the structures and energies of some alkenes and cycloalkanes by the force field method. *J. Am. Chem. Soc.*, **94**, 5734–5747

Anderson, N. S., Campbell, J. W., Harding, M. M., Rees, D. A. and Samual, J. W. B. (1969). X-ray diffraction studies of polysaccharide sulphates. *J. Mol. Biol.*, **45**, 85–99

Aspinall, G. O. (1987). Chemical modification and selective fragmentation of polysaccharides. *Acc. Chem. Res.*, **20**, 114–120

Atwood, J. L., Davies, J. E. D. and MacNicol, D. D. (1984). *Inclusion Compounds*, Volumes 2 and 3. Academic Press, New York

Barnes, J., Finney, J. L., Nicholas, J. D. and Quinn, J. E. (1979). Cooperative effects in simulated water. *Nature*, **282**, 459–464

Berendsen, H. J. C. (1988). Dynamic simulation as an essential tool in molecular modelling. *J. Comput. Mol. Design.*, **2**, 217–221

Berendsen, H. J. C., Postma, J. P. M., van Gunsteren, W. F., DiNola, A. and Haak, J. R. (1984). Molecular dynamics with coupling to an external bath. *J. Chem. Phys.*, **81**, 3684–3690

Berendsen, H. J. C., Postma, J. P. M., van Gunsteren, W. F. and Hermans, J. (1981). In Pullman, B. (Ed.), *Intermolecular Forces*. Reidel, Dordrecht, pp. 331–342

Berendsen, H. J. C., Grigera, J. R. and Straatsma, T. P. (1987). The missing term in effective pair potentials. *J. Phys. Chem.*, **91**, 6269–6271

Betzel, C., Saenger, W., Hingerty, B. E. and Brown, G. M. (1984). Circular and flip–flop hydrogen bonding in β-cyclodextrin undecahydrate: a neutron diffraction study. *J. Am. Chem. Soc.*, **106**, 7545–7556

Bock, K., Meldal, M., Bundle, D. R., Iversen, T., Garegg, P. J., Norberg, T., Lindberg, A. A. and Svenson, S. B. (1984). The conformation of salmonella O-antigenic polysaccharide chains of serogroups A, B and D1 predicted by semi-empirical, hard-sphere (HSEA) calculations. *Carbohydr. Res.*, **130**, 23–24

Brady, J. W. (1986). Molecular dynamics simulations of α-glucose. *J. Am. Chem. Soc.*, **108**, 8153–8160

Brady, J. W. (1987). Molecular dynamics simulations of β-D-glycopyranose. *Carbohydr. Res.*, **165**, 306–312

Brant, D. A., Tonelli, A. E. and Flory, P. J. (1969). The conformational statistics of random poly(lactic acid) chains. II. Theory. *Macromolecules*, **2**, 228–235

Burgen, A. S. V., Roberts, G. C. K. and Tute, M. S. (Eds.) (1986). *Molecular Graphics and Drug Design*, Topics in Molecular Pharmacology, Vol. 3. Elsevier, Amsterdam

Case, M. (1982). Electronic structure calculations using the X–α method. *Annu. Rev. Phys. Chem.*, **33**, 151–171

Ceccarelli, C., Jeffrey, G. A. and Taylor, R. (1981). Survey of O—H---O hydrogen bond geometries determined by neutron diffraction. *J. Mol. Struct.*, **70**, 255–271

Chacko, K. K. and Saenger, W. (1981). Topography of cyclodextrin inclusion complexes. Crystal and molecular structure of the cyclohexamylose 7.57 water complex, form III. Four and six membered circular hydrogen bonds. *J. Am. Chem. Soc.*, **103**, 1708–1714

Cramer, F., Saenger, W. and Spatz, H. C. (1967). Inclusion Compounds. XIX. The formation of inclusion compounds of α-cyclodextrin in aqueous solutions. Thermodynamics and kinetics. *J. Am. Chem. Soc.*, **89**, 14–20

Darsey, J. A., Voll, R. J., Younathan, E. S., and Koerner, T. A. W. (1988). *Ab initio* SCF energy calculations of the rotational orientation of each of the exocyclic groups of 6-*O*-methyl-β-D-tagatofuranose. *Carbohydr. Res.*, **173**, 197–204

Davies, D. B. and Christofides, J. C. (1987). Comparison of intramolecular hydrogen-bonding conformations of sucrose-containing oligosaccharides in solution and in the solid state. *Carbohydr. Res.*, **163**, 269–274

Diner, S., Malrieu, J. P. and Claverie, P. (1969). Localized bond orbitals and the correlation problem. I. The perturbation calculation of the ground state energy. *Theor. Chim. Acta*, **13**, 1–17

Fletcher, R. and Reeves, C. M. (1964). Function minimisation by conjugate gradients. *Comput. J.*, **7**, 149–154

Fujinaga, M., Gros, P. and van Gunsteren, W. F. (1989). Testing the method of crystallographic refinement with molecular dynamics. *J. Appl. Crystallogr.*, **22**, 1–8

Gelb, R. I., Schwartz, L. M., Johnson, R. F. and Laufer, D. A. (1979). The complexation chemistry of cyclohexamyloses. 4. Reactions of cyclohexamylose with formic, acetic and benzoic acids and their conjugate bases. *J. Am. Chem. Soc.*, **101**, 1869–1874

Goodall, D. M. and Norton, I. T. (1987). Polysaccharide conformations and kinetics. *Acc. Chem. Res.*, **20**, 59–65

van Gunsteren, W. F. and Karplus, M. (1982). Effect of constraints on the dynamics of macromolecules. *Macromolecules*, **15**, 1528–1544

van Gunsteren, W. F., Berendsen, H. J. C., Hermans, J., Hol, W. G. J. and Postma, J. P. M. (1983). Computer simulation of the dynamics of hydrated protein crystals and its comparison with X-ray data. *Proc. Natl. Acad. Sci. USA*, **80**, 4315–4319

Hamilton, J. A. and Chen, L. (1988). Crystal structure of an inclusion complex of β-cyclodextrin with racemic fenoprofen: direct evidence for chiral recognition. *J. Am. Chem. Soc.*, **110**, 5833–5841

Hingerty, B., Klar, B., Hardgrove, G. L., Betzel, C. and Saenger, W. (1984). Neutron diffraction of α-, β- and γ-cyclodextrins: hydrogen bonding patterns. *J. Biomol. Struct. Dyn.*, **2**, 249–260

Imberty, A. and Perez, S. (1988). A revisit to the three-dimensional structure of B-type starch. *Biopolymers*, **27**, 1205–1221

Jeffrey, G. A. and Maluszynska, H. (1986). A survey of the geometry of hydrogen bonds in the crystal structures of barbiturates, purines and pyrimidines. *J. Mol. Struct.*, **147**, 127–142

Jeffrey, G. A. and Mitra, J. (1984). Three center (bifurcated) hydrogen bonding in the crystal structures of amino acids. *J. Am. Chem. Soc.*, **106**, 5546–5553

Jeffrey, G. A., Pople, J. A. and Radom, L. (1974). The application of *ab initio* molecular orbital theory to structural moieties of carbohydrates. *Carbohydr. Res.*, **38**, 81–95

Karplus, M. (1987). The prediction and analysis of mutant structures. *Protein Eng* , **1**, 35–44

Kennedy, J. F. (Ed.) (1988). *Carbohydrate Chemistry*. Clarendon, Oxford

Klar, B., Hingerty, B. and Saenger, W. (1980). Topography of cyclodextrin inclusion complexes. XII. Hydrogen bonding in the crystal structure of α-cyclodextrin hexahydrate: the use of a multicounter detector in neutron diffraction. *Acta Crystallogr. B*, **36**, 1154–1165

Koehler, J. E. H. (1985). Gaussian 82 calculations on cyclic and chain-like hydrogen-bonded systems. *Diploma Thesis*, FU Berlin

Koehler, J. E. H. and Saenger, W. (1987). *Ab initio* calculations and many body analysis of the water tetramer. *Proceedings of the 35th Bunsen Colloquium, Marburg*. Springer, Berlin

Koehler, J. E. H., Lesyng, B. and Saenger, W. (1987a). Cooperative effects in extended hydrogen bonded systems involving OH-groups. *Ab initio* studies of the cyclic water tetramer. *J. Comput. Chem.*, **8**, 1090–1098

Koehler, J. E. H., Saenger, W. and van Gunsteren, W. F. (1987b). A molecular dynamics simulation of crystalline α-cyclodextrin hexahydrate. *Eur. Biophys. J.*, **15**, 197–210

Koehler, J. E. H., Saenger, W. and van Gunsteren, W. F. (1987c). Molecular dynamics simulation of crystalline β-cyclodextrin dodecahydrate at 293 K and 120 K. *Eur. Biophys. J.*, **15**, 211–224

Koehler, J. E. H., Saenger, W. and van Gunsteren, W. F. (1988a). Molecular dynamics simulation of crystalline β-cyclodextrin: the flip–flop hydrogen-bonding phenomenon. *Eur. Biophys. J.*, **16**, 153–168

Koehler, J. E. H., Saenger, W. and van Gunsteren, W. F. (1988b). Conformational differences between α-cyclodextrin in aqueous solution and in crystalline form: a molecular dynamics study. *J. Mol. Biol.*, **203**, 241–250

Koehler, J. E. H., Saenger, W. and van Gunsteren, W. F. (1988c). On the occurrence of three center hydrogen bonds in cyclodextrins in crystalline form and in aqueous solution: comparison of neutron diffraction and molecular dynamics results. *J. Biomol. Struct. Dyn.*, **6**, 181–198

Kollman, P. A. (1985). Theory of complex molecular interactions: computer graphics, distance geometry, molecular mechanics and quantum mechanics. *Acc. Chem. Res.*, **18**, 105–111

Kollman, P. A. and Allen, L. C. (1972). The theory of the hydrogen bond. *Chem. Rev.*, **72**, 283–303

Kuriyan, J., Petsko, G. A., Levy, R. M. and Karplus, M. (1986). Effect of anisotropy and anharmonicity on protein crystallographic refinement. An evaluation by molecular dynamics. *J. Mol. Biol.*, **190**, 227–254

Lemieux, R. U. (1978). Human blood groups and carbohydrate chemistry. Haworth Memorial Lecture 1978. *Chem. Soc. Rev.*, **7**, 423–452

Lesyng, B. and Saenger, W. (1981). Theoretical investigations on circular and chain-like hydrogen bonded structures found in two crystal forms of α-cyclodextrin hexahydrate. *Biochim. Biophys. Acta*, **678**, 408–413

Lindner, K. and Saenger, W. (1980). Crystal structure of the γ-cyclodextrin. *Biochem. Biophys. Res. Commun.*, **92**, 933–938

Lindner, K. and Saenger, W. (1982). Topography of cyclod. Inclusion complexes. XVI: Cyclic systems of hydrogen bonds. Structure of α-cyclodextrin hexahydrate, Form II. Comparison with Form I. *Acta Cryst.*, **B38**, 203–210

Lindon, J. C., Vinter, J. G., Lifely, M. R. and Moreno, C. (1984). Conformational and dynamic differences between *N. meningitidis* serogroup B and C polysaccharides, using NMR spectroscopy and molecular mechanics calculations. *Carbohydr. Res.*, **133**, 59–74

Linert, W., Han, L. and Lukovits, I. (1989). The use of the isokinetic relationship and molecular mechanics to investigate molecular interactions in inclusion complexes of cyclodextrins. *Chem. Phys.*, **139**, 441–455

Lipkind, G. M., Verovsky, V. E. and Kochetkov, N. K. (1984). Conformational states of cellobiose and maltose in solution: a comparison of calculated and experimental data. *Carbohydr. Res.*, **133**, 1–13

Manor, P. C. and Saenger, W. (1974). Topography of cyclodextrin inclusion complexes. III. Crystal and molecular structure of cyclohexamylose hexahydrate, the $(H_2O)_2$ inclusion complex. *J. Am. Chem. Soc.*, **96**, 3630–3639

Melberg, S. and Rasmussen, K. (1978). A simplified force field for glycopyranose. *Acta Chem. Scand. A*, **32**, 187–188

Melberg, S. and Rasmussen, K. (1979). Conformations of disaccharides by empirical force field calculations: part 1, β-maltose. *Carbohydr. Res.*, **69**, 27–38

Menger, F. M. and Sherrod, M. J. (1988). Docking calculations of ferrocene complexation with cyclodextrins. *J. Am. Chem. Soc.*, **110**, 8606–8611

Mezei, M., Beveridge, D. L., Berman, H. M., Goodfellow, J. M., Finney, J. L. and Neidle, S. (1983). Monte Carlo studies on water in the dCpG/proflavin crystal hydrate. *J. Biomol. Struct. Dyn.*, **1**, 287–297

Neszmelyi, A. and Hollo, J. (1989). Some aspects of the structure of starch-a 3-D molecular modelling approach. *Starch*, **41**, 1–3

Pechenaya, V. I. and Serikov, A. A. (1988). Study of furanose ring flexibility in polynucleotide chains using Raman spectra analysis. *Biopolymers*, **27**, 1817–1829

Plueckthun, A. and Skerra, A. (1989). Expression of functional antibody Fv and Fab fragments in *E. coli. Methods Enzymol.*, **178**, 497–515

Postma, J. P. M. (1985). *Ph.D. Thesis*, University of Groningen

Ramachandran, G. N. and Sasisekharan, V. (1968). Conformation of polypeptides and proteins. *Adv. Protein Chem.*, **23**, 283–438

Ramamurthy, V. and Eaton, D. F. (1988). Photochemistry and photophysics within cyclodextrin cavities. *Acc. Chem. Res.*, **21**, 300–306

Rees, D. A. and Scott, W. E. (1971). Polysaccharide conformation. Part VI. Computer model building for linear and branched pyranoglycans. Correlations with biological function. Preliminary assessment of inter-residue forces in aqueous solution. *J. Chem. Soc. B*, 469–479

Rees, D. A. and Skerrett, R. J. (1968). Conformational analysis of cellobiose, cellulose and xylan. *Carbohydr. Res.*, **7**, 334–348

Rees, D. A. and Smith, P. J. C. (1975). Polysaccharide conformation. Part IX. Monte Carlo calculation of conformational energies for disaccharides and comparison with experiment. *J. Chem. Soc., Perkin Trans. II*, 836–840

Rees, D. A. and Thom, D. (1977). Polysaccharide conformation. Part X. Solvent and temperature effects on the optical rotation and conformation of model carbohydrates. *J. Chem. Soc., Perkin Trans. II*, 191–201

Richards, W. G. (1989). *Computer-aided Molecular Design*. IBC Technical Services Ltd, London

Ruch, E. (1972). Algebraic aspects of the chirality phenomenon in chemistry. *Acc. Chem. Res.*, **5**, 49–56

Saenger, W. (1979). Circular hydrogen bonds. *Nature*, **279**, 343–344

Saenger, W. (1980). Cyclodextrin inclusion compounds in research and industry. *Angew. Chem., Int. Edn. Engl.*, **19**, 344–362

Saenger, W. (1987). Structure and dynamics of water surrounding biomolecules. *Annu. Rev. Biophys. Chem.*, **16**, 93–114

Saenger, W., Betzel, C., Hingerty, B. and Brown, G. M. (1982). Flip–flop hydrogen bonding in a partially disordered system. *Nature*, **296**, 581–583

Saenger, W., Hunter, W. N. and Kennard, O. (1986). DNA conformation is determined by economics in the hydration of phosphate groups. *Nature*, **324**, 385–388

Schneider, H. J., Kramer, R., Simova, S. and Schneider, U. (1988). Solvent and salt effects on binding constants of organic substrates in macrocyclic host compounds. A general equation measuring hydrophobic binding contributions. *J. Am. Chem. Soc.*, **110**, 6442–6448

Skerra, A. and Plueckthun, A. (1988). Assembly of a functional immunoglobuline F_v-fragment in *E. coli*. *Science*, **240**, 1038–1041

Stillinger, F. H. and David, C. W. (1978). Polarization model for water and its ionic dissociation products. *J. Chem. Phys.*, **69**, 1473–1484

Struempel, M. and Luger, P. (1988). Conformational calculations for the α and β anomer of 2,3,4-trio-*O*-acetyl-D-arabinopyranosyl azide. *Carbohydr. Res.*, **180**, 129–135

Szejtli, J. (1977). Einige Anwendungsmoeglichkeiten der Cyclodextrine in der Arzneimittel-industrie. *Staerke*, **29**, 26–33

Szejtli, J. (1983). Physiological effects of cyclodextrins on plants. *Starch*, **35**, 433–438

Szejtli, J. (1984). Highly soluble β-cyclodextrin derivates. *Starch*, **36**, 429–432

Szejtli, J., Fenyvesi, E. and Zsadon, B. (1978). Cyclodextrin polymere. *Staerke*, **30**, 127–131

Tajmir-Riahi, H. A. (1988). Carbohydrate adducts with zinc-group-metal ions. Interaction of β-D-fructose with Zn(II), Cd(II), and Hg(II) cations, and the effects of metal-ion coordination on the sugar isomer binding. *Carbohydr. Res.*, **172**, 1–10

Talashek, T. A. and Brant, D. A. (1987). The influence of sidechains on the calculated dimensions of the related bacterial polysaccharides. *Carbohydr. Res.*, **160**, 303–316

Vovelle, F., Goodfellow, J. M., Savage, H. F. J., Barnes, O. and Finney, J. L. (1985). Solvent structure in vitamin B12 coenzyme crystals. *Eur. Biophys. J.*, **11**, 225–237

Wiberg, K. B. and Murcko, M. A. (1988). Rotational barriers. 2. Energies of alkane rotamers. An examination of *gauche* interactions. *J. Am. Chem. Soc.*, **110**, 8029–8038

Wolfe, S., Whangbo, M. H. and Mitchell, D. J. (1979). On the magnitudes and origins of the 'anomeric effects', 'exo-anomeric effects', 'reverse anomeric effects', and C–X and C–Y bond lengths in XCH_2YH molecules. *Carbohydr. Res.*, **69**, 1–26

Yalpani, M. (Ed.) (1988). *Industry Polysaccharides. Genetic Engineering, Structure/Property Relations Applications*. Elsevier, Amsterdam

Zabel, V., Saenger, W. and Mason, S. A. (1986). A neutron diffraction study of the hydrogen bonding in β-cyclodextrin undecahydrate at 120 K: from dynamic flip–flops to static homodromic chains. *J. Am. Chem. Soc.*, **108**, 3664–3673

4
Prediction of the Three-dimensional Structures of Ribonucleic Acids: from tRNA to 16S Ribosomal RNA

Arun Malhotra, Robert K.-Z. Tan and Stephen C. Harvey

INTRODUCTION

Naturally occurring RNAs can display either long double-helical structures, as in viral dsRNA, or a globular structure with short double-helical domains connected by single-stranded stretches (Saenger, 1984). Globular RNAs are involved in transcription and translation, and form important components of ribosomes, spliceosomes, ribozymes, and several viruses. Reliable prediction of the three-dimensional structures of RNA chains in these molecular structures can give important insights into their functional behaviour.

Although a lot of effort has been put into the prediction of the three-dimensional structures of proteins, relatively little research has been carried out on the prediction of the three-dimensional structure of RNAs. There are two reasons, both related to RNA secondary structure prediction, that give cause for optimism in developing methods for three-dimensional modelling of an RNA from its primary structure. First, the algorithms for predicting RNA secondary structure are more reliable for nucleic acids than they are for proteins. Whereas the latter have reliabilities in the range of 60–70% for the prediction of α helical regions and β sheet regions, secondary structure of RNAs can be predicted with much greater reliability, particularly when phylogenetic sequence comparison methods are used. The second reason is that, unlike proteins, the information on secondary structure in nucleic acids can yield long-distance structural information. For proteins, even when the secondary structure can be predicted accurately, the α helices are only local structures and there are no methods for predicting which regions of β sheet will associate with one another and whether they will align in a parallel or an antiparallel direction. RNA secondary structure, on the other hand, generally contains information on close physical contact between regions which are distant from one another in the primary sequence. Nucleic acids are also made up of a limited number of very similar components, unlike the amino acids which are more diverse, making RNA chains easier to model than proteins.

RNA secondary structure prediction is an important step in predicting the three-dimensional form of an RNA molecule. Most RNA secondary structure predictions use one or a combination of two approaches—a thermodynamic approach using free energy methods, or a phylogenetic sequence comparison approach (Turner *et al.*, 1988). The free energy approach, pioneered by Tinoco *et al.* (1971), is based on minimizing the free energy change associated with nearest-neighbour interactions, and focuses on finding the lowest free energy structure. Secondary structures near the minimum free energy can also now be generated (Zuker, 1989). The comparative sequence analysis method (Fox and Woese, 1975; Woese *et al.*, 1983) looks for structural features which are conserved during evolution, assuming that RNA molecules with similar functions in different organisms have similar secondary structure. These methods have been very successful with the ribosomal RNAs (Noller, 1984).

Apart from the secondary structure, experimental data establishing proximity between disparate regions of the RNA chain can provide crucial clues for predicting three-dimensional structure. Such data can be provided by cross-linking, footprinting and chemical accessibility experiments. Other data such as helix and base stacking patterns are also important structural clues.

Several RNA chains exist as RNA–protein complexes, as in the small and large ribosomal subunits, and the different small nuclear ribonucleoprotein particles (snRNPs). These proteins serve as a scaffold, holding different regions of the RNA in a definite relationship to one another. For such systems, data on the associated proteins—shapes, locations, as well as RNA–protein and protein–protein contacts—are essential to model the RNA conformation.

X-ray crystallography methods can provide direct evidence for three-dimensional modelling of RNA chains. tRNA is the only RNA chain for which successful X-ray crystal structures (Kim *et al.*, 1973; Hingerty *et al.*, 1978; Sussman *et al.*, 1978; Schevitz *et al.*, 1979; Moras *et al.*, 1980; Woo *et al.*, 1980) are available, although crystals of ribosomal RNA are now available (Yonath *et al.*, 1987). The large size of ribosomal particles, which makes detailed X-ray analysis difficult, allows direct investigation by electron microscopy, which has yielded shapes and three-dimensional outlines of the 30S and the 50S subunits (Stöffler and Stöffler-Meilicke, 1986). Immunoelectron microscopy has also been used to localize proteins and RNA epitopes on the surface of ribosomal particles (Stöffler and Stöffler-Meilicke, 1986). Electron microscopy techniques are, however, useful only for looking at the surface features of large RNAs.

More detailed prediction of the three-dimensional structure of RNAs other than tRNA has relied on manually built mechanical or computer graphic models. Mechanical models have a long and colourful history in nucleic acid research, starting from the classic models of Watson and Crick (1953) to the current models of the 16S ribosomal RNA (Brimacombe *et al.*, 1988). More recently, manually manipulated computer graphics techniques have emerged, and are being used to model the 16S rRNA in the *Escherichia coli* 30S ribosomal subunit (Nagano *et al.*, 1988; Schuler and Brimacombe, 1988; Stern *et al.*, 1988c). Detailed atomic models are also being built using interactive assembly and modification of refined RNA fragments, as was used to model 5S RNA of spinach chloroplast and of *Xenopus laevis* somatic and oocyte 5S RNA by Westhof *et al.* (1989).

These manually built models incorporate a variety of data—secondary structure, cross-linking and base-protection patterns, protein maps, and results from immunoelectron microscopic analysis. Apart from being time consuming and laborious to develop, manually built models have several drawbacks. They present only a single conformation out of all the possible folding patterns that can satisfy experimental data. With such models it is difficult to judge inconsistencies and conflicts in data quantitatively, as well as to take the inherent errors in experimental data into account. The model building process can also introduce biases into mechanically constructed models, because the interior regions of such models are not easy to manipulate.

In this chapter we introduce a computer-assisted procedure for folding large RNA chains into three-dimensional conformations consistent with their secondary structure and other known experimental constraints (Malhotra *et al.*, 1990). Figure 4.1 shows how a randomly oriented RNA chain can be folded, as additional constraints are applied, to generate an acceptable conformation. Our procedure folds randomly oriented structures using molecular mechanics techniques and can be used to build a variety of conformations that satisfy the available structural data.

Molecular mechanics uses energy minimization and molecular dynamics to model a system of atoms (McCammon and Harvey, 1987). In a traditional molecular mechanics all-atom model each atom is represented by a point mass. Potential energy functions are used to mimic bonds, bond angles and torsions, van der Waals interactions, electrostatics and other constraints and forces among these atoms. These empirical potential functions can be used to calculate the potential energy surface, which gives the energy of the system as a function of the atomic coordinates. Energy minimization is used to search for the global minimum energy structure in such models.

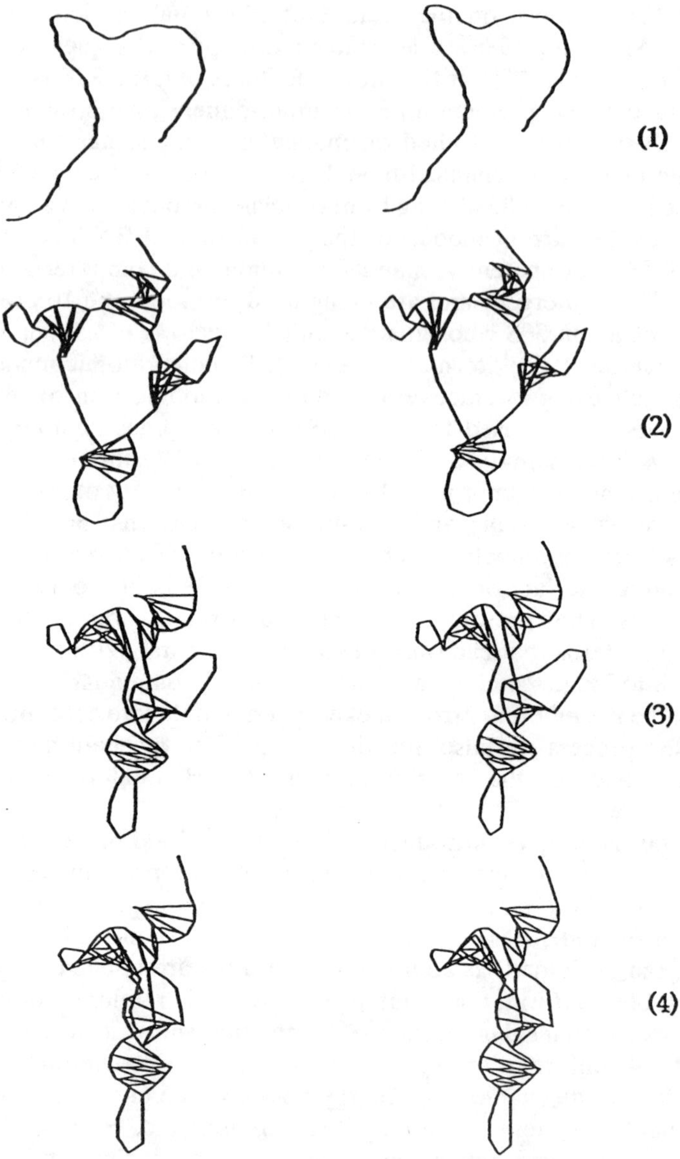

Figure 4.1 Generation of a three-dimensional model for yeast tRNA^{Phe} from a random chain. The above sequence shows a random chain (1) being folded to greater refinement as additional constraints are applied: (2) secondary structure only, (3) secondary structure and correct helix stacking, and (4) secondary structure with correct helix stacking and contact between the D and T loops

The forces on the atoms in such a model are directly related to the first derivative of the potential surface with respect to the atom positions. These forces can be used in Newton's equations of motion to calculate how the atomic positions change with time—molecular dynamics. Molecular dynamics simulates intramolecular motions, and can provide information about the dynamical properties of the system. A variety of techniques are used for optimization of macromolecular models (McCammon and Harvey, 1987; Brünger, 1990). A combination of energy minimization and molecular dynamics, called simulated annealing, has been shown to be particularly powerful for optimization of such systems. Our procedure uses energy minimization for modelling structures of ribonucleic acids.

All-atom molecular modelling, although straightforward, is currently not possible for very large molecules. Energy minimization is computationally demanding for macromolecules, and at present only molecules with up to a few thousand atoms can be modelled in full detail. Apart from the computational requirements, the complexity of the potential energy surface increases with the number of atoms, hindering any search for the global minimum.

The disadvantages of all-atom modelling for macromolecules can be partially overcome by reduced representations (also called succinct models; Tan and Harvey, 1990) where pseudoatoms are used to represent a set of related atoms with a generally invariant structure such as a nucleotide or regular helix. Reduced representations were used in early simulations of protein folding (Levitt and Warshel, 1975; Levitt, 1976), although this approach was later abandoned because of the difficulties of treating long-range interactions which are so important for proteins. Nucleic acids are more amenable to reduced representations, as the major forces stabilizing RNA structure are base stacking and hydrogen bonding, both of which are short-range interactions. Our procedure uses reduced representations with pseudoatoms and force constants chosen to mimic a real RNA chain.

AN AUTOMATED RNA FOLDING PROCEDURE

We use several levels of reduced representations for the RNA chain in our models. The most intuitive of these is to represent each nucleotide as a single pseudoatom located at the phosphate atom (P-atom) of the nucleotide. We call this the all-phosphate representation, and models using such pseudoatoms for the entire RNA chain are called all-phosphate models. A coarser reduced representation (the 1H representation) involves using a single large pseudoatom (the helix or H-atom) to represent each of the double-stranded helical stems, while P-atoms are used for the single-stranded regions. We also have an intermediate level of representation using five (or more) pseudoatoms per helix (the 5H representation),

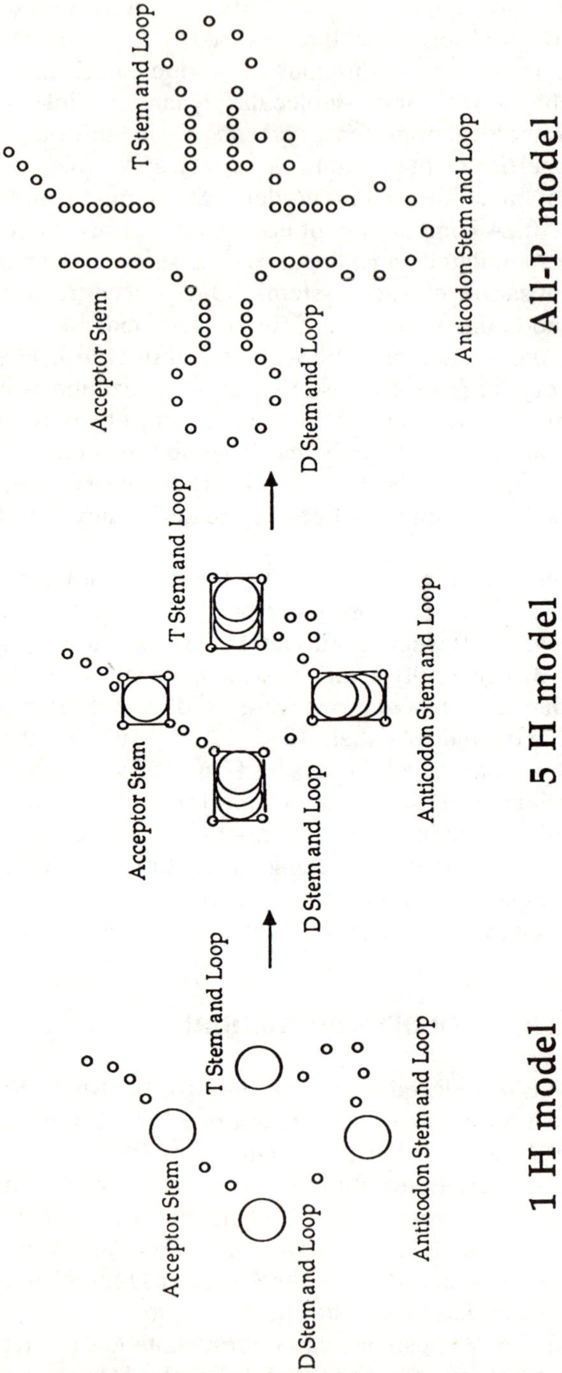

Figure 4.2 Schematic representation of the 1H, 5H and all-P models for $tRNA^{Phe}$

with large central atom(s) for space filling and the four corner P-atoms in the helix to orient the helix correctly. Figure 4.2 shows the three model levels for $tRNA^{Phe}$. For RNA systems with proteins such as the ribosomal RNA, the proteins are modelled as spheres with point masses, with the radii calculated from their anhydrous molecular weights (Richards, 1977).

Different secondary structure motifs—helices, bulges, loops, and single-stranded regions—are imposed on the pseudoatoms using bonds, angles and torsions, similar to the covalent bonds, angles and torsions used in traditional all-atom molecular mechanics. Helical regions of the RNA chain are assumed to have an ideal A-RNA conformation (Arnott *et al.*, 1976), the most common form of RNA double helix at low ionic strength (Saenger, 1984). The double helix is imposed using bonds between the hydrogen-bonded basepairs and neighbouring nucleotides. Angles are specified along the helical backbone and the correct chirality is maintained using improper torsions. Nucleotides in single-stranded regions are linked by bonds. Nucleotides in loop (hairpin and internal) regions are also linked by single bonds, although angles and torsions are applied to hairpin loops during initial minimization to mimic the helical characteristics of these loops. Helices with bulges are modelled as regular helices with the bulged nucleotide(s) attached to the neighbouring bases by suitable bonds and angles. Collinear or stacked helices are represented by extending helical constraints to connect the two (or more) helical regions involved. Other experimental data such as cross-links and close contacts (RNA–RNA or RNA–protein) are incorporated in the models by the use of appropriate bonds.

We use harmonic potential functions for bonds, angles and torsions. These potential functions are

$$E_{\beta_i} = k_{\beta_i}(\beta_i - \beta_{i0})^2 \qquad \text{for bonds}$$
$$E_{\alpha_i} = k_{\alpha_i}(\alpha_i - \alpha_{i0})^2 \qquad \text{for angles}$$
$$E_{\tau_i} = k_{\tau_i}(\tau_i - \tau_{i0})^2 \qquad \text{for torsions}$$

where E_{β_i} denotes the energy of the *i*th bond, k_{β_i}, k_{α_i} and k_{τ_i} are the force constants for the *i*th bond, angle and torsion respectively, α_i, β_i, and τ_i are the *i*th angle, bond length and torsion, and α_{i0}, β_{i0} and τ_{i0} are the corresponding equilibrium values. Harmonic functions are used in our models because they are easy to minimize and have a single unique minimum. A harmonic bond potential function is also equivalent to a Gaussian distribution of bond length β_i about β_{i0}, with a standard deviation equal to $RT/2k_{\beta_i}$. Force constants can thus be chosen to reflect the standard deviation expected about the equilibrium bond lengths, angles or torsions. In our models, the force constants are chosen to mimic variability in the $tRNA^{Phe}$ crystal coordinates (Hingerty *et al.*, 1978), one of the few well-characterized RNA structures. For example, an examination of the helical region of $tRNA^{Phe}$ shows that the separation of phosphate atoms in a given basepair has a standard deviation of about 1 Å (1 nm = 10 Å). The force constant of the harmonic potential function for such bonds is thus 1.2456 kJ mol^{-1} Å^{-2} (0.298 kcal mol^{-1} Å^{-2}). Similar analysis is used for other structural features—a standard deviation of 0.6 Å for the phosphate–phosphate distances in single-stranded regions and a standard deviation of 0.2 rad for angles and torsions in helices. Force constants for bonds representing experimental data are chosen to reflect experimental uncertainty. For cross-links, a standard deviation of 2 Å is used in the cross-linking length. RNA–protein close contacts are represented by bonds of 5 Å with a standard deviation of 2 Å.

Non-bond interactions are used to exclude volume occupied by the pseudoatoms. We use harmonic terms for non-bond interactions:

$$\begin{aligned} E_{\gamma_{ij}} &= k_{\gamma_{ij}}(r_{ij} - r_{ij0})^2 \qquad && \text{if } r_{ij} \leq r_{ij0} \\ &= 0 && \text{if } r_{ij} > r_{ij0} \end{aligned}$$

where $E_{\gamma_{ij}}$ is the non-bond interaction energy between atoms i and j, $k_{\gamma_{ij}}$ is the non-bond force constant for the atom pair ij, r_{ij} is the distance between atoms i and j, and r_{ij0} is the minimum distance allowed between the two atoms (usually the sum of their radii). The minimum separation of phosphate atoms in single-stranded regions of tRNA^{Phe} is used to get an exclusion radius of 5 Å for the P-pseudoatoms. Suitable exclusion radii are used for other pseudoatoms.

For small RNA chains (~100 nucleotides), such as tRNAs, modelling can be started directly at the all-phosphate representation. Figure 4.1 shows an all-phosphate representation of tRNA^{Phe}, and illustrates folding of such a model into a three-dimensional structure. For larger RNA molecules, it is often necessary to use several different levels of reduced representation, beginning with a low-resolution 1H model of the RNA chain. This model is refined using energy minimization to determine the gross structural features, and the resulting folded chain is then extrapolated to a higher level model with more detail. Extrapolation between models is done using superposition of ideal helices on the helical pseudoatoms (the radii of helical pseudoatoms in the 1H and the 5H models are chosen to accommodate an ideal helix). This allows modelling to proceed from a randomly oriented chain at the 1H representation, to the 5H representation, and then to the all-phosphate representation, to get a satisfactory conformation for the RNA chain backbone (Figure 4.3 shows this procedure of refinement and extrapolation for tRNA). In principle, this procedure can be used to extrapolate up to an all-atom model of the RNA chain; all-atom extrapolation rules are obvious for helical regions but are yet to be determined for other regions.

Starting structures are generated using random walk chains for the pseudoatoms in the model. The direction at each step of the walk is varied randomly between zero and a maximum specified angle, and the length of

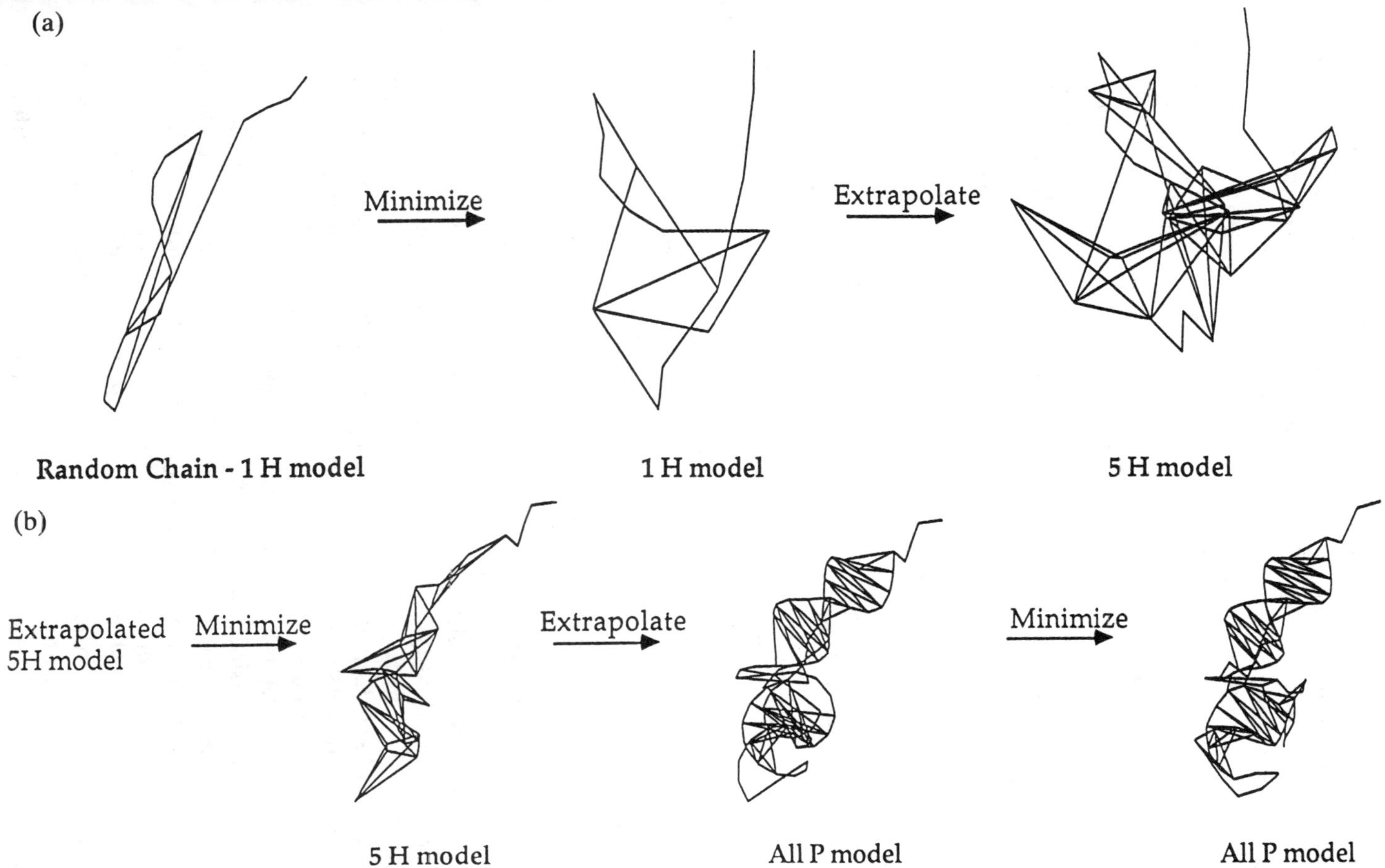

Figure 4.3 Example of a random chain folded into tRNA by successive minimization and extrapolation with increasing detail at each step: (a) from a random 1H model to a 5H model; (b) from a 5H model to an all-P model of tRNA

each step is adjusted according to the length of the bond connecting two neighbouring pseudoatoms. Several chains, with different random seeds and maximum angles, are used to get a variety of starting coordinates.

Energy minimization, using steepest descent and conjugate gradient methods, is used to refine the models. Energy minimization is done with *yammp*, an in-house molecular mechanics package (Tan and Harvey, 1989). All models are initially minimized with soft non-bond interactions (Levitt, 1983) designed to give an energy of 4.18 kJ mol^{-1} (1 kcal mol^{-1}) for an overlap of 1 Å between two pseudoatoms. This low overlap penalty permits the chain to pass through itself, allowing tangles in the starting structure to be resolved. The resulting structures are then further minimized with stiffer non-bond force constants (to give an energy of 418 kJ mol^{-1} (100 kcal mol^{-1}) for a 1 Å overlap between two pseudoatoms). Not all randomly generated initial structures can be untangled—these typically minimize to higher energies, and are excluded from further refinement and analysis.

This RNA folding procedure has been tested with $tRNA^{Phe}$, and gives results consistent with the known structure of tRNAs. These results are discussed below. Applications of the procedure to the *E. coli* 16S rRNA are discussed later in the chapter.

tRNA MODELLING: A TEST FOR THE FOLDING PROCEDURE

Transfer RNAs are the best-understood RNA molecules. Good quality X-ray crystal structures are available for yeast $tRNA^{Phe}$ from orthorhombic (Sussman *et al.*, 1978) and monoclinic (Hingerty *et al.*, 1978) crystal forms, as well as for yeast $tRNA^{Asp}$ (Moras *et al.*, 1980), and the initiator $tRNA^{fMet}$ from yeast (Schevitz *et al.*, 1979) and *E. coli* (Woo *et al.*, 1980). All transfer RNAs, which vary in length between 75 and 90 nucleotides, follow the cloverleaf-type secondary structure as shown in Figure 4.4 for yeast $tRNA^{Phe}$. Even before X-ray crystallographic structures became available in 1973 (Kim *et al.*, 1973), about seven different models (summarized by Levitt, 1969) had been proposed for the tertiary folding of tRNA, based on chemical and physical data available at that time.

To test the RNA folding procedure, we looked at different structural models of tRNA using only secondary structure data known before the crystal structure became available. The RNA folding procedure can be used to test different constraints—any model with cross-link, stacking or other data that are incompatible with the secondary structure and other tertiary constraints will minimize to a conformation that has higher energy than a conformation where all the constraints are satisfied. This property was used to investigate the different possible stackings of the acceptor stem, D stem, anticodon stem and the T stem in $tRNA^{Phe}$. Four different

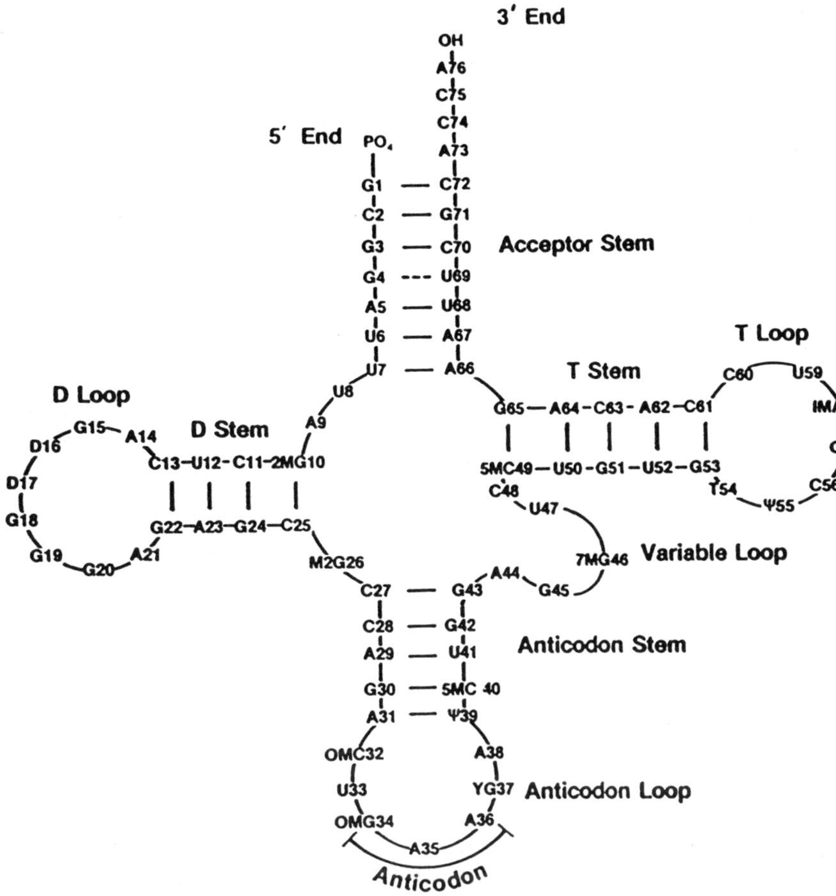

Figure 4.4 Cloverleaf representation of the primary and secondary structure of $tRNA^{Phe}$

random chains were minimized using constraints that required different patterns of helix stacking and loop–loop interactions. The RNA chains were modelled at an all-phosphate level of detail. The energies of the minimized structures are tabulated in Table 4.1. As can be seen from this table, models which imposed stacking of the acceptor stem on the anticodon stem, and the D stem on the T stem, minimized to energies several orders of magnitude higher than other stacking schemes, indicating that such a stacking has several unsatisfied constraints (mostly steric clashes) and is not acceptable in a three-dimensional model of tRNA. On the other hand, the tRNA secondary structure is flexible enough to accommodate several other stacking and loop–loop interaction schemes. The crystal structure of $tRNA^{Phe}$ (Hingerty *et al.*, 1978) has shown that the D stem is stacked on the anticodon stem, the acceptor stem is stacked on

Table 4.1 Energies of minimized structures

Helix stacking	*Loop–loop interaction*	*Chain number*	*Energy (arbitrary units)*
Acceptor stem on D stem, and anticodon stem on T stem	D–anticodon loops	1	$<10^{-7}$
		2	$<10^{-7}$
		3	$<10^{-7}$
		4	$<10^{-7}$
	D–T loops	1	$<10^{-7}$
		2	$<10^{-7}$
		3	$<10^{-7}$
		4	$<10^{-7}$
Acceptor stem on anticodon stem, and D stem on T stem	D–anticodon loops	1	14.73
		2	4.09
		3	0.1749
		4	8.979
	Anticodon–T loops	1	26.6
		2	7.656
		3	1.25
		4	15.92
Acceptor stem on T stem, and D stem on anticodon stem	D–T loops	1	$<10^{-7}$
		2	$<10^{-7}$
		3	$<10^{-7}$
		4	$<10^{-7}$
	Anticodon–T loops	1	$<10^{-7}$
		2	$<10^{-7}$
		3	$<10^{-7}$
		4	$<10^{-7}$

the T stem, and the D and T loops are in close proximity, an arrangement that is clearly acceptable in our models.

It would seem that our procedure, although good at picking out unacceptable conformations, does not do a good job of identifying the correct stacking and loop interaction scheme. This is because the quality of the conformations generated by the folding procedure is dependent on the number and quality of tertiary constraints that are added to the model. Clearly, if additional experimentally observed constraints are introduced into the model (for example, radius of gyration, cross-links between specific nucleotides on two stems), a smaller number of arrangements would have been acceptable.

As a further test of the procedure, it was used to model $tRNA^{Phe}$ with only the secondary structure information (Figure 4.4), the stacking of the D stem on the anticodon stem, stacking of the acceptor stem on the T stem, and required close contact between the D and T loops (secondary structure and three additional tertiary constraints). Using different initial random walk chains, several possible conformations consistent with the constraints listed above were generated. A few of these are shown in Figure 4.5. As can be seen from the different conformations shown, the anticodon and the

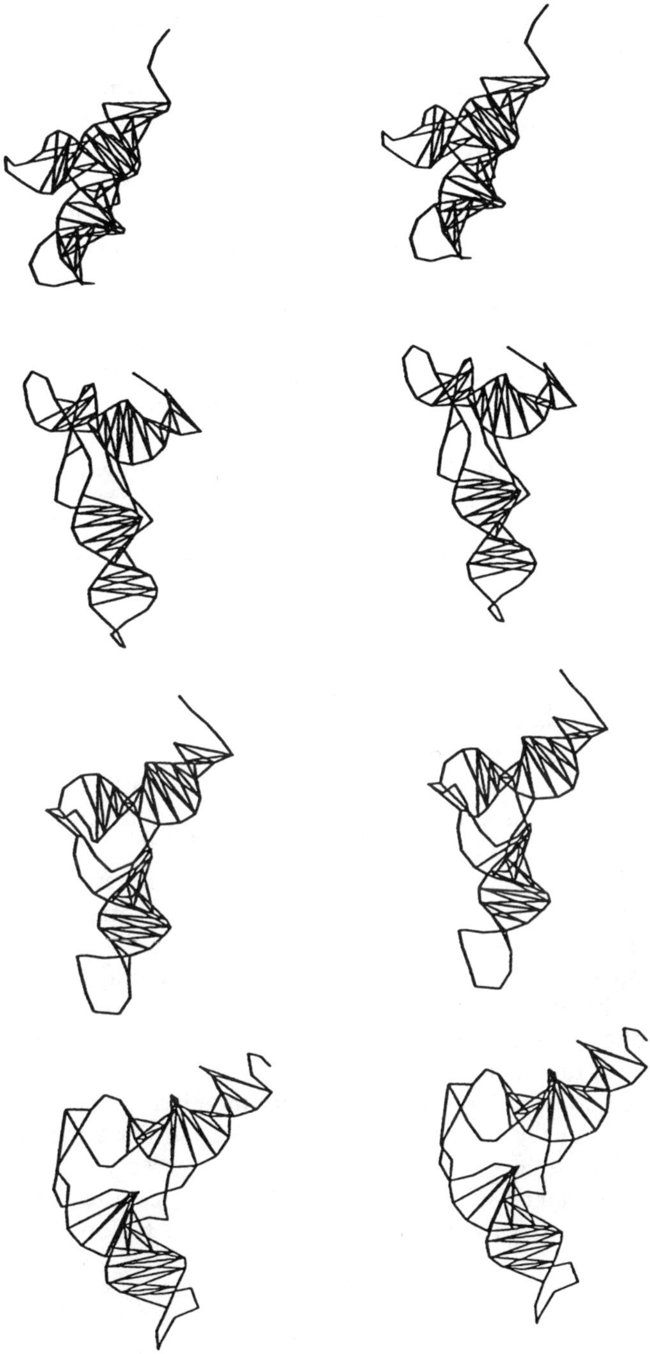

Figure 4.5 Some of the final refined structures generated for $tRNA^{Phe}$ by specifying the secondary structure, correct helix stacking, and D–T loop interactions

acceptor arms are each well predicted, and they can take different stereochemically acceptable conformations with respect to each other. This is an indication that the region between the two arms is not well defined with the constraints used. The form of the two-armed tRNA is, however, clearly evident even with the very small number of constraints used.

The procedure was also tested by taking the X-ray crystal structure coordinates (Hingerty *et al.*, 1978) as the initial coordinates for folding. As expected these coordinates had a much lower initial energy when compared with a random chain, and rapidly minimized to a conformation similar to the crystal structure with an RMS deviation of 2.28 Å between the folded and X-ray crystal coordinates. Unlike the tRNA crystal structure which has near-ideal helices and non-collinear stacking of the D stem and the anticodon stem, the folding procedure imposes ideal A-RNA helices, and perfect collinear stacking of both pairs of stems. This is responsible for the small difference between the folded and the crystallographic conformations, even when ideal initial coordinates are used for folding.

These tests illustrate several important aspects of our folding procedure. Firstly, for most sets of constraints, several different acceptable conformations can be generated (as seen in Figure 4.5 for tRNA). The procedure does not give a single 'correct' three-dimensional structure, unless a very large number of constraints ($3N$-6 constraints, where N is the number of atoms in the model) are specified. For most systems the number of constraints is much less than this ideal number, giving rise to an underdetermined problem which has several possible solutions. The procedure should instead be used to get a feel for the range and variability among the different possible models.

Secondly, the folding procedure imposes ideal structures for helices and stacking, and is not currently capable of modelling variations at the level of nucleotides. This also depends on the coarseness of the representation—for example, the 1H model is accurate only to the level of individual helices. With future improvements in this procedure to extrapolate the all-phosphate models to greater detail, it will be possible to model variations at the nucleotide level. Internal loops, bulges and other single-stranded regions are greatly underdetermined in our models, as not much is known about the structures of these regions in RNA molecules (Tinoco *et al.*, 1987).

Thirdly, as can be seen from Table 4.1, this procedure can be a powerful tool to test alternative tertiary constraints, and to look for inconsistent data.

MODELLING RIBOSOMAL RNA: 16S RNA IN THE *E. coli* 30S SUBUNIT

The ribosome is a complex ribonucleoprotein system made up of three large rRNAs and 52 (in *E. coli*) proteins responsible for translation of genetic information on mRNAs into polypeptides during protein synthesis. Because of the importance of this polymerase, it has been the subject of intense research for the past three decades. An important key to understanding the ribosomal mechanism lies in its three-dimensional structure. The ribosome is made up of two subunits—the small subunit (30S in prokaryotes, and 40S in eukaryotes), and the large subunit (50S in prokaryotes and 60S in eukaryotes). The smaller subunit of the *E. coli* ribosome, the 30S subunit, is made up of 21 proteins (S1–S21) and an RNA chain (16S) with 1542 nucleotides. With the availability of a complete map of the protein positions in the small subunit of the *E. coli* ribosome (Capel *et al.*, 1988), and a large body of cross-linking (summarized by Brimacombe *et al.*, 1988), and footprinting (summarized by Stern *et al.*, 1989) data, it is now possible to construct plausible three-dimensional models of the 30S subunit (Brimacombe, 1988). Models for the structure of the *E. coli* 16S RNA chain in the protein framework of the 30S subunit have been built manually, using mechanical models (Brimacombe *et al.*, 1988) and interactive computer graphics (Nagano *et al.*, 1988; Stern *et al.*, 1988c). An earlier mechanical model of the 16S RNA was constructed by Expert-Bezancon and Wollenzien (1985).

Data Used

The secondary structure of the 16S RNA (Figure 4.6) for our modelling studies is taken from Stern *et al.* (1988c). This secondary structure is based on the 16S secondary structure derived by comparative sequence analysis (Gutell *et al.*, 1985). Additional tertiary constraints, provided by cross-linking data and footprinting studies from several different laboratories, are also used in our models. Some of these are shown in Figure 4.6.

Cross-linking data provide a good indication of regions lying close to each other in the 30S subunit. Earlier cross-linking studies on the 16S RNA used inactive 30S subunits (Wollenzien and Cantor, 1982) and naked 16S RNA (Expert-Bezancon *et al.*, 1983; Thompson and Hearst, 1983; Wollenzien *et al.*, 1985). It has been suggested that these cross-links, which formed the basis of the three-dimensional model of 16S RNA by Expert-Bezancon and Wollenzien (1985), do not fully reflect tertiary RNA arrangement in the intact, active 30S subunit (Stiege *et al.*, 1986). Newer cross-links using UV irradiation with intact 30S subunits have been reported, both *in vivo* (Stiege *et al.*, 1986) and *in vitro* (Atmadja and

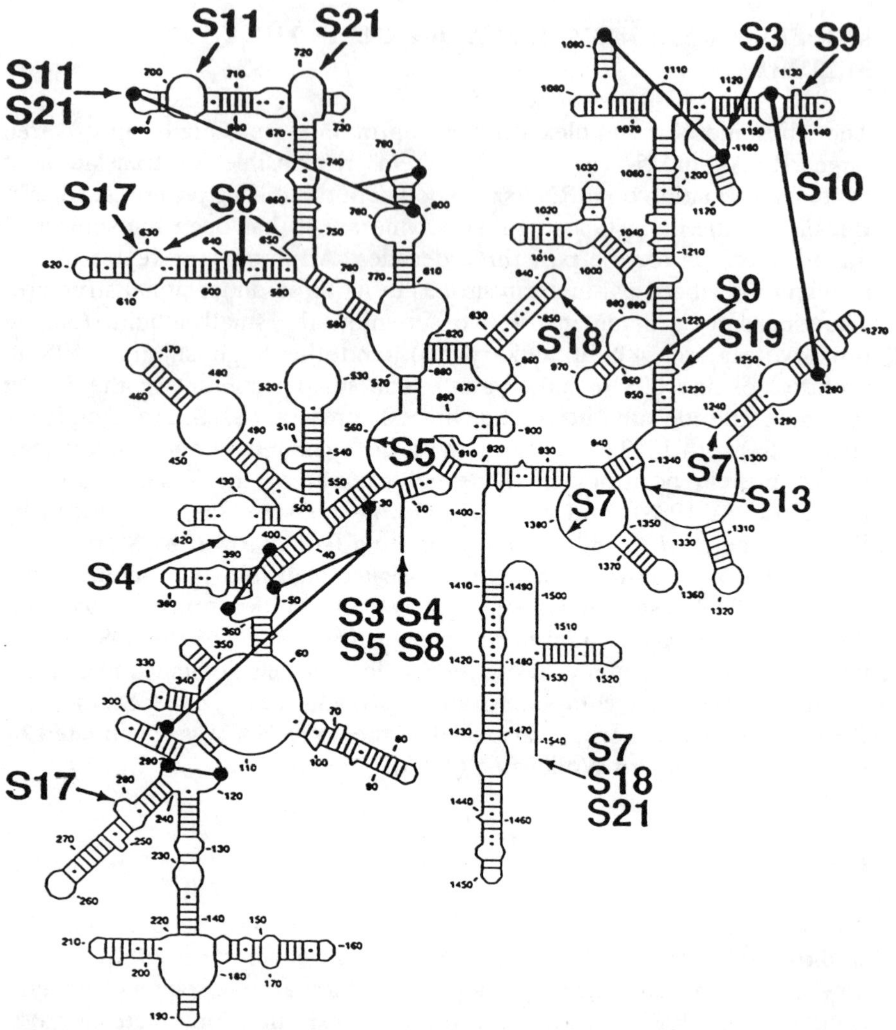

Figure 4.6 Secondary structure and RNA–RNA and protein–RNA cross-linking data for the *E. coli* 16S ribosomal RNA. Bold solid lines show RNA–RNA cross-links. Arrows show locations of RNA cross-links to the small subunit proteins (S1–S21). (Figure reproduced here from Stern *et al.* (1988c) with permission)

Brimacombe, 1985). Nitrogen mustard (*bis*(2-chloroethyl)-methylamine) has also been used as an *in vitro* cross-linker for intact 30S subunits (Atmadja *et al.*, 1986). Eight tertiary RNA–RNA cross-links identified by these studies (summarized by Brimacombe *et al.*, 1988) are used in our models. 23 RNA–protein cross-links are included in our models, based on the RNA–protein cross-links summarized by Brimacombe *et al.* (1988). These are all cross-links identified in active 30S subunits using methyl *p*-azidophenyl acetimidate (Osswald *et al.*, 1987), UV radiation (Zwieb and Brimacombe, 1979), nitrogen mustard (Greuer *et al.*, 1987) and 2-iminothiolane (Wower and Brimacombe, 1983; Kyriatsoulis *et al.*, 1986). Of the reported cross-links to the 3′ and the 5′ termini of the 16S RNA, only two are used in our model—cross-links between protein S5 and the 5′ terminus, and between S18 and the 3′ terminus. These RNA–RNA and RNA–protein cross-links are shown in Figure 4.6. The anomalous cross-link between S13 and positions 189–191 (Osswald *et al.*, 1987) is included in our models; the role of this cross-link will be discussed later.

Footprinting data can also provide important tertiary clues about the interaction between the ribosomal proteins and the 16S RNA. These data identify nucleotides in the 16S RNA that show enhanced or reduced reactivities to chemical and enzymatic probes on the addition of different 30S subunit ribosomal proteins. Primer extension methods have been used to examine the interaction of the ribosomal proteins S16, S17, S20 (Stern *et al.*, 1988a), S5, S6, S11, S18, S21 (Stern *et al.*, 1988b), S7, S9, S19 (Powers *et al.*, 1988a), S2, S3, S10, S13, S14 (Powers *et al.*, 1988b), S6, S8, S15 and S18 (Svensson *et al.*, 1988) with 16S RNA. These studies (summarized in Stern *et al.*, 1988c; Stern *et al.*, 1989) used chemical probes kethoxal, dimethyl sulphate, 1-cyclohexyl-*S*-(2-morpholinoethyl)-carbodiimide metho-*p*-toluenesulphonate and nuclease probes RNase (ribonuclease) T_1, RNase A and RNase V_1. Regions of the RNA chain strongly protected from chemical attack by a protein are assumed to be in close contact with that protein in our model. For each cluster of two or more (neighbouring or basepaired) nucleotides strongly protected from chemical attack, a single close contact is included in our model, yielding a total of 62 RNA–protein close contacts. For protein S14, which has a single protected nucleotide (A1362), a close contact between the protein and the nucleotide is also included in our model as no other tertiary data exist for S14. Footprinting results showing enhanced reactivities were not included in our model, as these are thought to be due to induced conformational changes and are unreliable indicators of protein location (Stern *et al.*, 1988c). Protection data with nuclease probes are not included in our model.

The 21 ribosomal proteins are modelled as spheres with radii taken as the anhydrous radii calculated from their molecular weights (Wittmann-Liebold, 1986). The proteins were tethered with bonds to the positions reported in the neutron scattering map of the small subunit (Capel *et al.*, 1988). Force constants for these bonds are chosen to mimic the uncertainty in protein positions as reported by Capel *et al.* (1988). The protein positions obtained by neutron scattering are based on distances between 93 protein pairs, each with a different experimental error, giving rise to an asymmetrical covariance matrix. However, force constants in our models, which are based on experimental error, are uniform in all directions. To mimic the asymmetry in the protein positions, all the 93 individual protein–protein distances used to generate the neutron scattering map are included as bonds (with their reported individual experimental uncertainty) in our models.

Refinement Procedures

Starting structures are generated using random walk chains for the 485 pseudoatoms in the 1H model of the 16S RNA. Pseudoatoms representing the 21 proteins are given the coordinates reported by Capel *et al.* (1988) and the geometric centre of the RNA chain is superimposed on that of the proteins. Several chains, with different random seeds and maximum angles, are used to get a variety of starting coordinates. Five different starting structures were minimized to get refined 1H models. To study the role of footprinting data in defining the minimized structures the 1H refinement procedures were repeated without the footprinting data for the five chains.

Of the five 1H models, three have been further extrapolated, and refined to an all-phosphate representation of the 16S RNA. One typical sequence of increasing refinement is shown in Figure 4.7, which shows refined 1H, 5H and all-phosphate model starting from a randomly oriented chain. The 5H model is made up of 840 pseudoatoms; the all-phosphate model has 1563 (1542 nucleotides and 21 proteins) pseudoatoms. Two refined all-phosphate models are shown in Figure 4.8. Refinement of a model at each level of detail—1H, 5H and all-P—takes about two hours of Cray X/MP CPU time.

MODELLING RESULTS

The 1H model refinement brings different parts of the random starting structure into correct relative positions. Further refinement, at the 5H and all-phosphate representation, mostly changes the orientation of different parts in the chain with no large changes in the overall shape and form of the model refined at the 1H level, as can be seen in Figure 4.7. Thus an analysis of different random chains refined at the 1H level can yield sufficient information about the positions of different parts of the 16S RNA chain in the minimized conformation.

The 16S RNA chain is not evenly constrained in our model, as the complexity of the secondary structure is not uniform and the experimental footprinting and cross-linking data are unevenly spread through the chain. Thus, some parts of the RNA chain can be better characterized than others. An analysis of the variability between the different refined 1H models can be used to identify ill-defined and well-characterized regions of

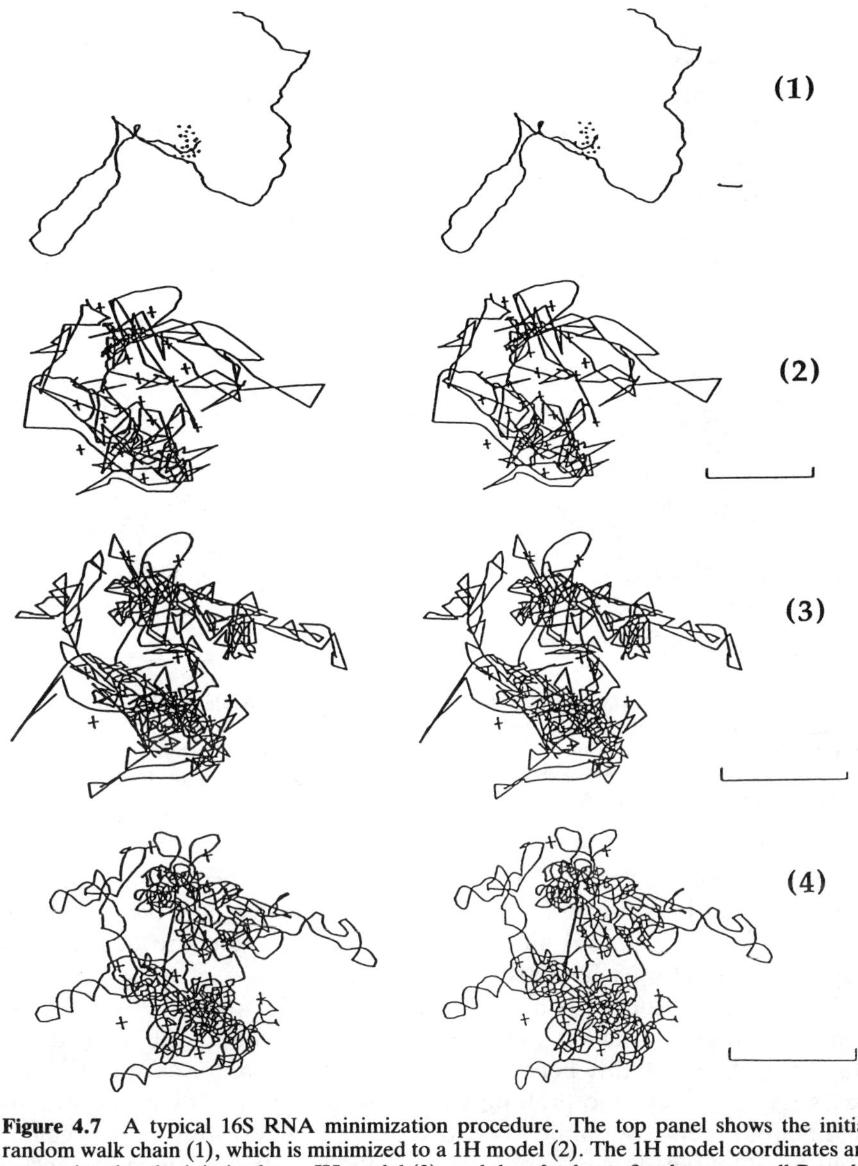

Figure 4.7 A typical 16S RNA minimization procedure. The top panel shows the initial random walk chain (1), which is minimized to a 1H model (2). The 1H model coordinates are extrapolated and minimized to a 5H model (3), and then further refined to get an all-P model (4). Stereo view shown is from the solvent interface (also called the cytoplasmic side) with the same orientation as Figure 18 of Stern *et al.* (1988c), and Figure 16 of Capel *et al.* (1988). The scale bars represent 100 Å. The small subunit protein centres are marked by crosses (protein numbering is shown in Figure 4.10)

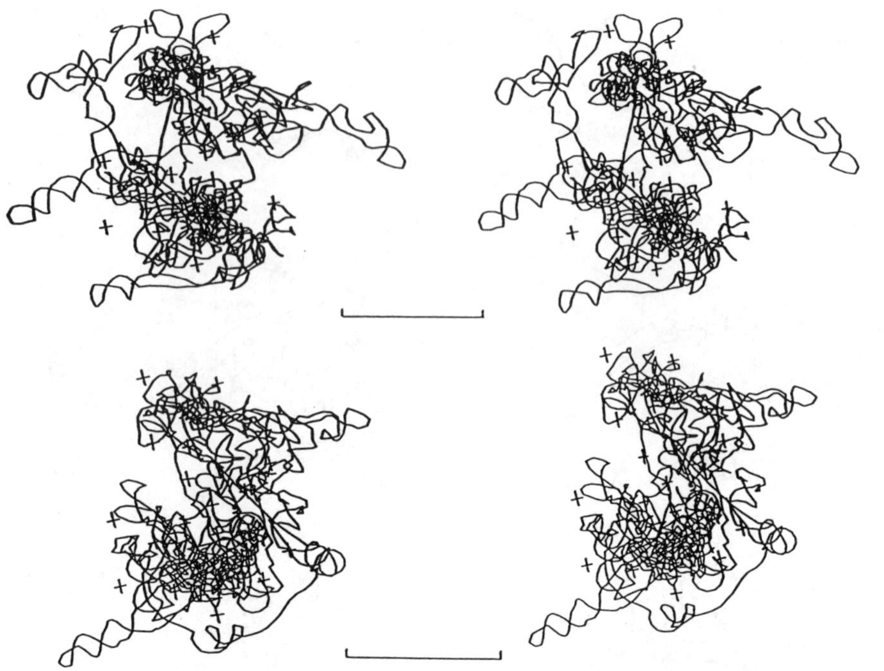

Figure 4.8 Two *E. coli* 16S RNA models refined to an all-phosphate resolution. The small subunit protein centres are marked by crosses. The parts of RNA that lie outside of the ribosome correspond to regions where there are no cross-linking or footprinting data to constrain the RNA chain. Stereo view shown is from the solvent interface (see Figure 4.7). The scale bars represent 100 Å

the RNA chain. Variability between the different conformations generated was examined by superposing the refined 1H models (only the 21 protein positions are superposed) and then calculating the standard deviation of the coordinates for each nucleotide in the models.

Results of such an analysis are shown in Figure 4.9 for the five refined 1H models. These five models were refined with and without the footprinting data, and the variability between the models is shown for each of the two cases. As expected, footprinting data provide crucial tertiary constraints and help to reduce variability between the different conformations generated by our procedure. This analysis identifies four ill-defined regions in the 16S RNA which require more cross-linking or footprinting data to be properly positioned in a three-dimensional model of the 30S subunit. These regions are the 59–180 region, including the 66–103 stem and the 144–178 stem, the 437–497 stem and internal loop, the 997–1044 region, and the penultimate stem (1409–1491). These regions of the RNA chain took widely different positions in the refined conformations and very little structural prediction can be made about them.

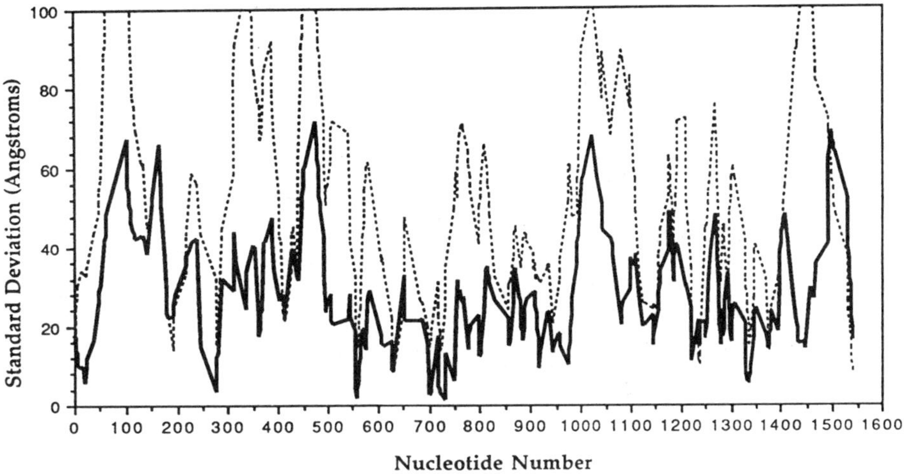

Figure 4.9 Variation (measured in terms of standard deviation) between coordinates of pseudoatoms in five refined 1H models of the *E. coli* 16S rRNA. This analysis is used to identify ill-defined regions in the 16S rRNA. Solid line refers to variation between 1H models minimized with footprinting data. Dotted line shows variations between 1H models minimized in absence of footprinting data

All the 1H conformations share several characteristics which can yield important clues to the three-dimensional structure of the 30S subunit. All our models have physical dimensions similar to the 30S subunit dimensions observed using electron microscopy—230 Å high, 140 Å wide, and 115 Å thick as reported by Lake *et al.* (1974). We also see a distinct separation of the 16S RNA into two major structural units. The positions of several regions of the RNA chain—especially the pseudoknot region and the central domain region—are similar in all the conformations generated so far, indicating that the data are sufficient to define these features relatively well.

The separate domains, as seen in the two refined all-phosphate models, are shown in Figures 4.10, 4.11, 4.12, and 4.13, illustrating the similarity and variability between our models. As shown in Figure 4.10, when viewed from the cytoplasmic or the solvent side, the pseudoknot region of the 16S RNA lies in the lower part of the subunit extending from above protein S16 to S4 and S5. Figure 4.11 shows the 5′ domain (37–546) region, which extends mostly from below S3 and S20 to S17 on the 50S interface side. The anomalous S13 and 189–191 cross-link (Brimacombe *et al.*, 1988), which is included in our model, causes part of the 5′ domain to fold up towards S13. The orientation of that part of the 5′ domain—nucleotides 136–227—is not well defined in our models. The central domain (557–918) is seen in the region of proteins S6, S21 and S8, and folds compactly into the bottom half of the 30S subunit in all our models (Figure 4.12). This part of the RNA

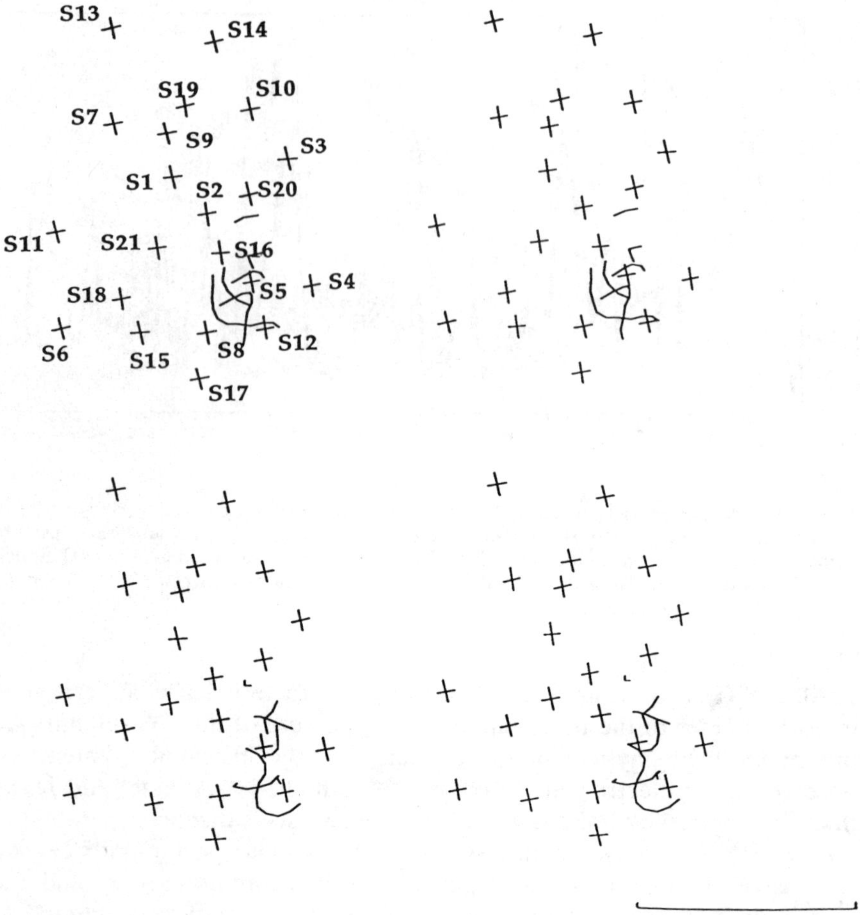

Figure 4.10 Pseudoknot regions of the two refined 16S rRNA models shown in Figure 4.8. RNA backbone is shown for residues 9–13, 17–25, 27–37, 547–556, and 916–918. The small subunit protein centres are marked by crosses. The scale bar represents 100 Å

chain is relatively well-defined. The top part or head of the 30S subunit, in our 1H models, is almost exclusively made up of the 3′ domain (920–1396), as can be seen in Figure 4.13.

Protein Positions and Cross-links in our Models

The 21 protein positions display an RMS deviation of 4–6 Å, when our 1H models are compared with the neutron scattering map. This is well within the range of standard errors in the protein map (Capel *et al.*, 1988). The largest displacement and variation of the protein positions, when com-

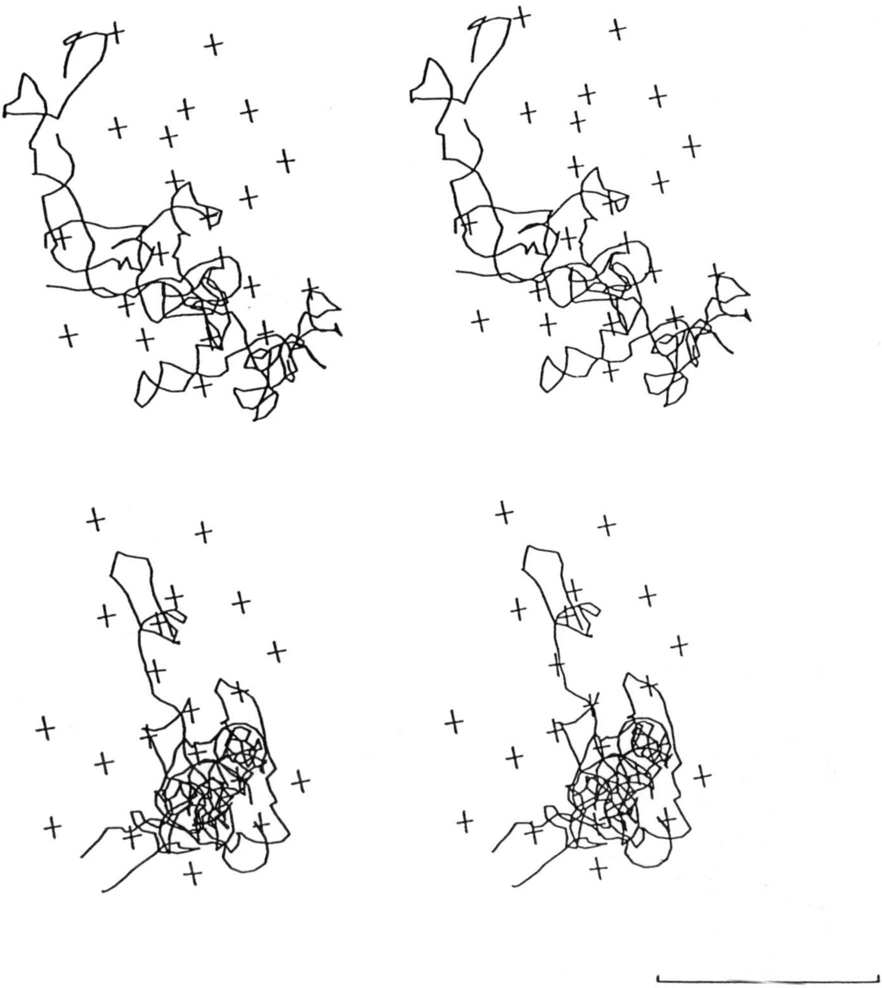

Figure 4.11 The well-defined regions of the 5′ domain in the two refined 16S rRNA models shown in Figure 4.8. RNA backbone is shown for residues 38–65, 104–143, 179–436, and 498–546. The small subunit protein centres are marked by crosses. The scale bar represents 100 Å

pared with the neutron scattering coordinates, was seen in protein S18. Bonds used to tether the proteins to the protein map coordinates are very weak (because of higher experimental uncertainty), compared with the secondary structure and cross-link bonds, and so individual proteins would be easily displaced in our model if their positions were inconsistent with other constraints. Thus, at the 1H level, the protein map data appear to have no serious inconsistencies. Cross-linking and base protection data, on the other hand, do have some inconsistencies. As described earlier, all the

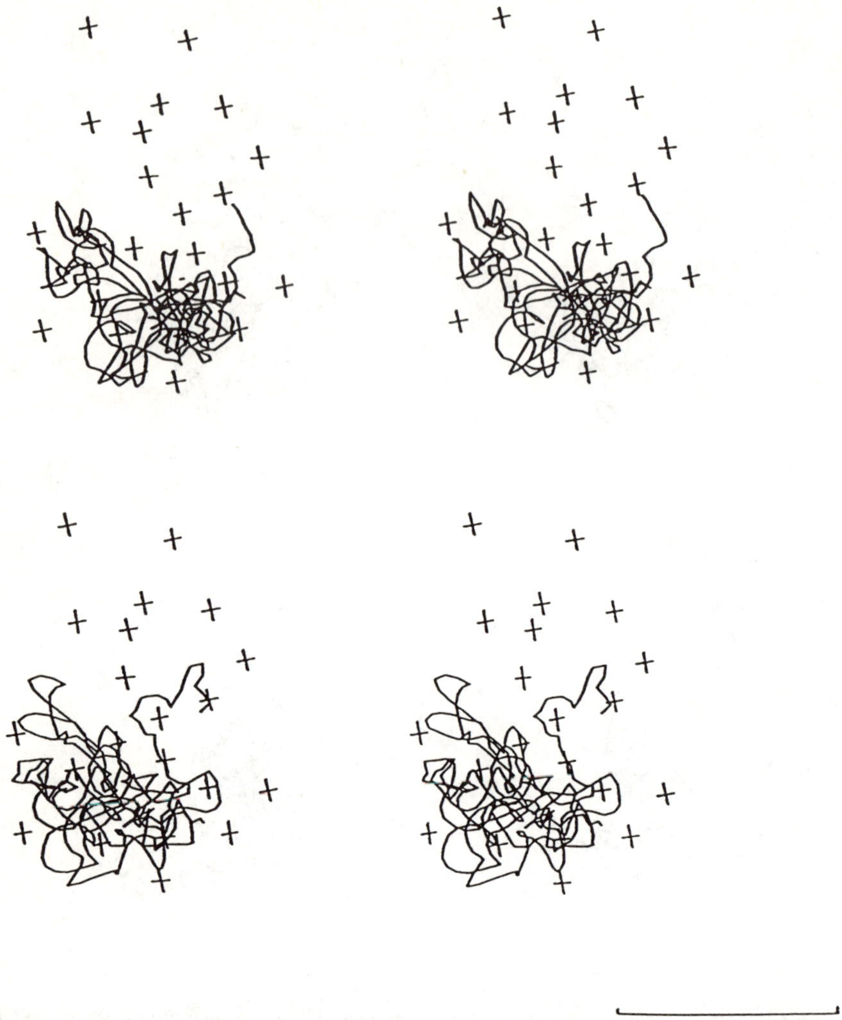

Figure 4.12 The well-defined regions of the central domain in the two refined 16S rRNA models shown in Figure 4.8. RNA backbone is shown for residues 558–915. The small subunit protein centres are marked by crosses. The scale bar represents 100 Å

reported cross-links to the 5′ and the 3′ termini cannot be simultaneously satisfied—we used only one link to each terminus. Several other cross-links and contacts are not satisfied completely in our models, and work is currently underway at our laboratory to identify cross-links which are inconsistent with other experimental data.

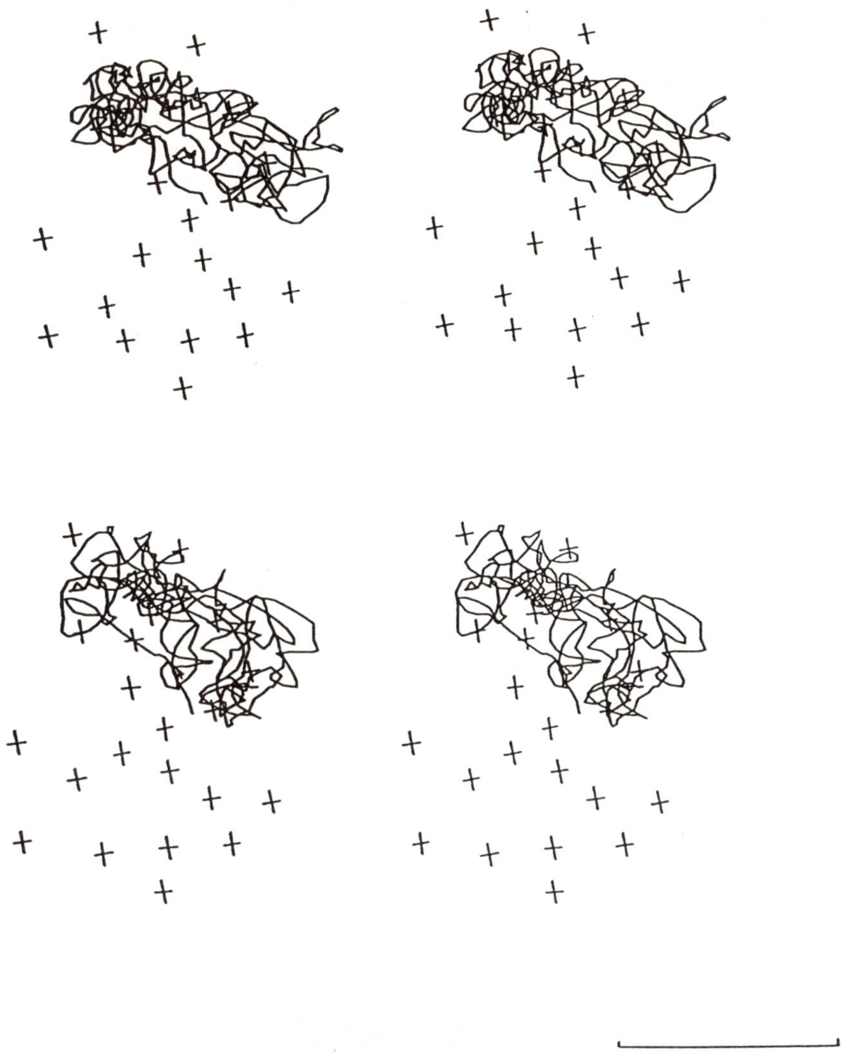

Figure 4.13 The well-defined regions of the 3′ domain in the two refined 16S rRNA models shown in Figure 4.8. RNA backbone is shown for residues 920–990 and 1046–1397. The small subunit protein centres are marked by crosses. The scale bar represents 100 Å

The S13–RNA Position 189–191 Cross-link

The protein S13 has been placed in the head of the 30S subunit by electron microscopy (Stöffler and Stöffler-Meilicke, 1986) and by the neutron scattering map of the small subunit proteins (Capel *et al.*, 1988). S13 is, however, implicated in cross-links with RNA sites in the 5′ domain, localized in the bottom of the small subunit (Osswald *et al.*, 1987). This, and other (summarized by Brimacombe *et al.*, 1988) anomalous observations about S13 may be due to different S13 positions in the ribosome (Capel *et al.*, 1988) or due to contact between different adjacent S30 subunits (Osswald *et al.*, 1987; Brimacombe *et al.*, 1988). The cross-link between S13 and 189–191 was included in our modelling, and is satisfied by folding of the 136–227 region of the 5′ domain up towards the top of the 30S subunit. Apart from this cross-link, no other tertiary data are available for this region of the 5′ domain. A single protection with S20 has been reported in this region of the 16S RNA by Stern *et al.* (1988a), but this did not meet the criteria of protection on two or more neighbouring nucleotides used in our model for tertiary contact. This lack of data is reflected in the large variability between different refined 1H models in this region of the 5′ domain (see Figure 4.9). Figure 4.14 shows a 1H resolution

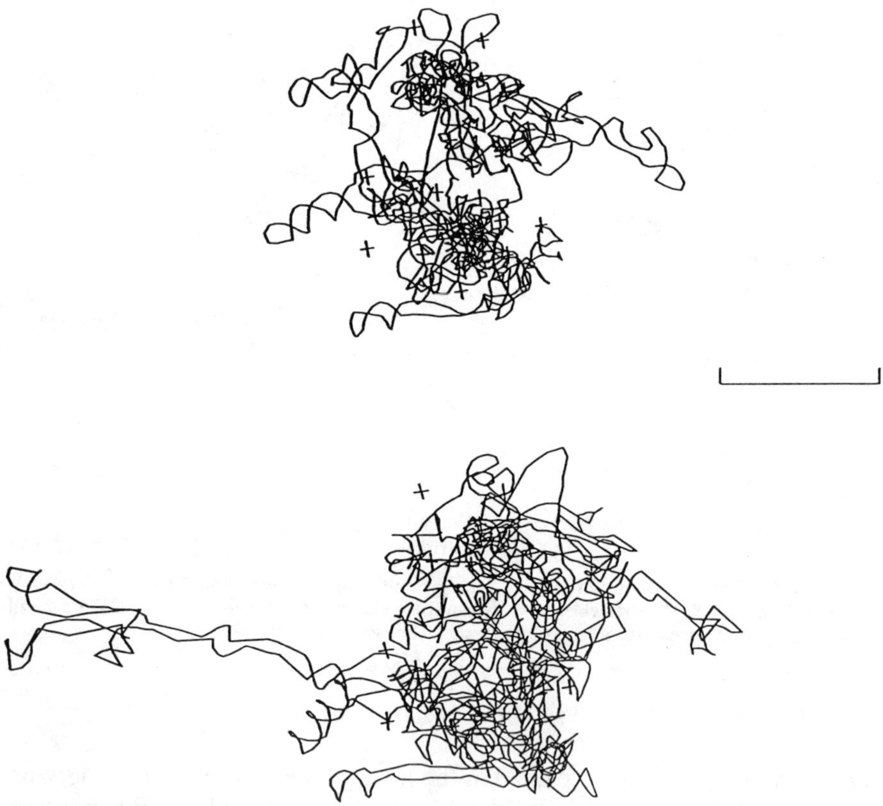

Figure 4.14 Random walk chain of the 16S rRNA refined with the S13–189 cross-link (top), and without the cross-link (refined only to 1H resolution) (bottom). View shown is from the solvent interface (same as Figure 4.7). The small subunit protein centres are marked by crosses. The scale bar represents 100 Å

model refined in the presence and the absence of the cross-link between S13 and 181–191, and illustrates how a single cross-link can alter the possible conformation of the RNA chain.

Comparisons with Other Current Models of the 16S RNA

The overall features of our model are similar to those of the interactive graphics model of 16S rRNA manually built by Stern *et al.* (1988c). Their model was based on data similar to those used for our study and it shows comparable positioning of the pseudoknot region and the three major domains of the RNA chain. The relative positionings of the different domains in our models are also comparable with the model of Brimacombe *et al.* (1988) which was based on RNA–RNA and RNA–protein cross-links. Because of the large variability which exists in our current models (of the order of 20 Å or more), no detailed comparison with these two models has been done so far. A similar level of uncertainty is also present in the manually built models of the 16S RNA (R. Brimacombe, personal communication).

CONCLUSIONS AND DISCUSSION

Secondary structure and other tertiary data for an RNA chain can usually be satisfied by several different conformations. The automated procedure to fold large RNA chains, using molecular mechanics techniques, described here is a powerful tool to investigate different possible three-dimensional conformations of RNA chains that satisfy all the available data. These conformations can then be examined for common three-dimensional themes.

Apart from yielding information about the three-dimensional structure of the RNA chain (for a large enough set of constraints), this procedure can also be used to identify ill-defined regions of an RNA chain, to check experimental data for consistency, and to suggest which experiments would provide the most information to narrow the range of acceptable models. Although our procedure omits the stereochemical intuition that goes into manually built models, the quantitative nature of our models, the ability to test different data rapidly and the determination of a range of models consistent with the data offers advantages not easily realized with mechanical modelling.

The use of succinct models for nucleic acids allows us to overcome size limitations of all-atom molecular mechanics, and makes it possible to model systems with hundreds or thousands of nucleotides. This chapter presented preliminary results on modelling of 16S RNA in the 30S subunit of *E. coli* ribosome. Many possible conformations can be generated to satisfy the currently available structural data on the 16S RNA; a few are presented here. Our current models are low resolution models that can be used to look at the relative positions of different regions in the RNA chain. Further refinement at the 5H and all-phosphate level of detail is currently underway to examine possible orientations of various helical regions. Similar succinct models are also being developed for other ribosomal RNAs and for supercoiled DNA (Tan and Harvey, 1989, 1990).

ACKNOWLEDGEMENTS

This work was supported by a grant from the National Science Foundation (DMB-87-06551). Computer time was provided by the Alabama Supercomputer Network. Additional computer support was provided by the Atherosclerosis Research Unit, University of Alabama at Birmingham. Superposition algorithms were implemented by Martin Jones.

REFERENCES

Arnott, S., Campbell Smith, P. J. and Chandrasekaran, R. (1976). In Fasman, G. D. (Ed.), *Handbook of Biochemistry and Molecular Biology*. Chemical Rubber Company Press, Cleveland, OH, pp. 411–422

Atmadja, J. and Brimacombe, R. (1985). Investigation of the tertiary folding of *Escherichia coli* 16S RNA by *in situ* intra-RNA crosslinking within 30S ribosomal subunits. *Nucleic Acids Res.*, **13**, 6919–6936

Atmadja, J., Stiege, W., Zobawa, M., Greuer, B., Osswald, M. and Brimacombe, R. (1986). The tertiary folding of *Escherichia coli* 16S RNA, as studied by *in situ* intra-RNA crosslinking of 30S ribosomal subunits with *bis*-(2-chloroethyl)-methylamine. *Nucleic Acids Res.*, **14**, 659–673

Brimacombe, R. (1988). The emerging three-dimensional structure and function of 16S ribosomal RNA. *Biochemistry*, **27**, 4207–4214

Brimacombe, R., Atmadja, J., Stiege, W. and Schuler, D. (1988). A detailed model of the three-dimensional structure of *Escherichia coli* 16S ribosomal RNA *in situ* in the 30S subunit. *J. Mol. Biol.*, **199**, 115–136

Brünger, A. T. (1990). Refinement of three-dimensional structures of proteins and nucleic acids. In Goodfellow, J. M. (Ed.), *Molecular Dynamics*. Macmillan, Basingstoke, Chap. 5 (this volume)

Capel, M. S., Kjeldgaard, M., Engelman, D. M. and Moore, P. B. (1988). Positions of S2, S13, S16, S17, S19 and S21 in the 30S ribosomal subunit of *Escherichia coli*. *J. Mol. Biol.*, **200**, 65–87

Expert-Bezancon, A. and Wollenzien, P. L. (1985). Three-dimensional arrangement of the *Escherichia coli* 16S ribosomal RNA. *J. Mol. Biol.*, **184**, 53–66

Expert-Bezancon, A., Milet, M. and Carbon, P. (1983). Precise localization of several covalent RNA–RNA cross-links in *Escherichia coli* 16S RNA. *Eur. J. Biochem.*, **136**, 267–274

Fox, G. E. and Woese, C. R. (1975). 5S RNA secondary structure. *Nature*, **256**, 505–507

Greuer, B., Osswald, M., Brimacombe, R. and Stöffler, G. (1987). RNA–protein cross-linking in *Escherichia coli* 30S ribosomal subunits: determination of sites on 16S RNA that are cross-linked to proteins S3, S4, S7, S9, S10, S11, S17, S18 and S21 by treatment with *bis*-(2-chloroethyl)-methylamine. *Nucleic Acids Res.*, **15**, 3241–3255

Gutell, R. R., Weiser, B., Woese, C. R. and Noller, H. F. (1985). Comparative anatomy of 16S-like ribosomal RNA. *Prog. Nucleic Acid Res. Mol. Biol.*, **32**, 155–216

Hingerty, B., Brown, R. S. and Jack, A. (1978). Further refinement of the structure of yeast $tRNA^{Phe}$. *J. Mol. Biol.*, **124**, 523–534

Kim, S.-H., Quigley, G. J., Suddath, F. L., McPherson, A., Sneden, D., Kim, J. J., Weinzierl, J. and Rich, A. (1973). Three-dimensional structure of yeast phenylalanine transfer RNA: folding of the polynucleotide chain. *Science*, **179**, 285–288

Kyriatsoulis, A., Maly, P., Greuer, B., Brimacombe, R., Stöffler, G., Frank, R. and Blocker, H. (1986). RNA–protein cross-linking in *Escherichia coli* ribosomal subunits: localization of sites on 16S RNA which are cross-linked to proteins S17 and S21 by treatment with 2-iminothiolane. *Nucleic Acids Res.*, **14**, 1171–1186

Lake, J. A., Sabatini, D. D. and Nonomura, Y. (1974). In Nomura, M., Tissieres, A. and Lengyel, P. (Eds.), *Ribosomes.* Cold Spring Harbor Laboratory Press, Cold Spring Harbor, NY, pp. 543–557

Levitt, M. (1969). Detailed molecular model for transfer ribonucleic acid. *Nature*, **224**, 759–763

Levitt, M. (1976). A simplified representation of protein conformations for rapid simulation of protein folding. *J. Mol. Biol.*, **104**, 59–107

Levitt, M. (1983). Protein folding by restrained energy minimization and molecular dynamics. *J. Mol. Biol.*, **170**, 723–764

Levitt, M. and Warshel, A. (1975). Computer simulation of protein folding. *Nature*, **253**, 694–698

Malhotra, A., Tan, R. K.-Z. and Harvey, S. C. (1990). Prediction of the three dimensional structure of *Escherichia coli* 30S ribosomal subunit—a molecular mechanics approach. *Proc. Natl. Acad. Sci. USA*, **87**, 1950–1954

McCammon, J. A. and Harvey, S. C. (1987). *Dynamics of Proteins and Nucleic Acids.* Cambridge University Press, Cambridge

Moras, D., Comarmond, M. B., Fischer, J., Weiss, R., Thierry, J. C., Ebel, J. P. and Giege, R. (1980). Crystal structure of yeast $tRNA^{Asp}$. *Nature*, **288**, 669–674

Nagano, K., Harel, M. and Takezawa, M. (1988). Prediction of three-dimensional structure of *Escherichia coli* ribosomal RNA. *J. Theor. Biol.*, **134**, 199–256

Noller, H. F. (1984). Structure of ribosomal RNA. *Annu. Rev. Biochem.*, **53**, 119–162

Osswald, M., Greuer, B., Brimacombe, R., Stöffler, G., Baumert, H. and Fasold, H. (1987). RNA–protein cross-linking in *Escherichia coli* 30S ribosomal subunits: determination of sites on 16S RNA that are cross-linked to proteins S3, S4, S5, S7, S8, S9, S11, S13, S19 and S21 by treatment with methyl *p*-azidophenyl acetimidate. *Nucleic Acids Res.*, **15**, 3221–3240

Powers, T., Changchien, L.-M., Craven, G. R. and Noller, H. F. (1988a). Probing the assembly of the 3′ major domain of 16S ribosomal RNA—quaternary interactions involving ribosomal proteins S7, S9 and S19. *J. Mol. Biol.*, **200**, 309–319

Powers, T., Stern, S., Changchien, L.-M. and Noller, H. F. (1988b). Probing the assembly of the 3′ major domain of 16S rRNA—interactions involving ribosomal proteins S2, S3, S10, S13 and S14. *J. Mol. Biol.*, **201**, 697–716

Richards, F. M. (1977). Area, volume, packing and protein structure. *Annu. Rev. Biophys. Bioeng.*, **6**, 151–176

Saenger, W. (1984). *Principles of Nucleic Acid Structure.* Springer, New York

Schevitz, R. W., Podjarny, A. D., Krishnamachari, N., Hughes, J. J., Sigler, P. B. and Sussman, J. L. (1979). Crystal structure of a eukaryotic initiator tRNA. *Nature*, **278**, 188–190

Schuler, D. and Brimacombe, R. (1988). The *Escherichia coli* 30S ribosomal subunit; an optimized three-dimensional fit between the ribosomal proteins and the 16S RNA. *EMBO J.*, **7**, 1509–1513

Stern, S., Changchien, L.-M., Craven, G. R. and Noller, H. F. (1988a). Interaction of proteins S16, S17 and S20 with 16S ribosomal RNA. *J. Mol. Biol.*, **200**, 291–299

Stern, S., Powers, T., Changchien, L.-M. and Noller, H. F. (1988b). Interaction of ribosomal proteins S5, S6, S11, S12, S18 and S21 with 16S rRNA. *J. Mol. Biol.*, **201**, 683–695

Stern, S., Weiser, B. and Noller, H. F. (1988c). Model for the three-dimensional folding of 16S ribosomal RNA. *J. Mol. Biol.*, **204**, 447–481

Stern, S., Powers, T., Changchien, L.-M. and Noller, H. F. (1989). RNA–protein interactions in 30S ribosomal subunits: folding and function of 16S rRNA. *Science*, **244**, 783–790

Stiege, W., Atmadja, J., Zobawa, M. and Brimacombe, R. (1986). Investigation of the tertiary folding of *Escherichia coli* ribosomal RNA by intra-RNA cross-linking *in vivo*. *J. Mol. Biol.*, **191**, 135–138

Stöffler, G. and Stöffler-Meilicke, M. (1986). In Hardesty, B. and Kramer, G. (Eds.), *Structure, Function, and Genetics of Ribosomes*. Springer, New York, pp. 28–46

Sussman, J. L., Holbrook, S. R., Wade Warrant, R., Church, G. M. and Kim, S.-H. (1978). Crystal structure of yeast phenylalanine tRNA. I. Crystallographic refinement. *J. Mol. Biol.*, **123**, 607–630

Svensson, P., Changchien, L.-M., Craven, G. R. and Noller, H. F. (1988). Interaction of ribosomal proteins, S6, S8, S15 and S18 with the central domain of 16S ribosomal RNA. *J. Mol. Biol.*, **200**, 301–308

Tan, R. K.-Z. and Harvey, S. C. (1989). Molecular mechanics model of supercoiled DNA. *J. Mol. Biol.*, **205**, 573–591

Tan, R. K.-Z. and Harvey, S. C. (1990). In Lavery, R. and Beveridge, D. (Eds.), *Theoretical Biochemistry and Molecular Biophysics*. Adenine Press, New York, in press

Thompson, J. F. and Hearst, J. E. (1983). Structure of *E. coli* 16S RNA elucidated by psoralen crosslinking. *Cell*, **32**, 1355–1365

Tinoco, I. Jr., Uhlenbeck, O. C. and Levine, M. D. (1971). Estimation of secondary structure in ribonucleic acids. *Nature*, **230**, 362–367

Tinoco, I. Jr., Davis, P. W., Hardin, C. C., Puglisi, J. D., Walker, G. T. and Wyatt, J. (1987). RNA structure from A to Z. *Cold Spring Harbor Symp. Quant. Biol.*, **52**, 135–146

Turner, D. H., Sugimoto, N. and Freier, S. M. (1988). RNA structure prediction. *Annu. Rev. Biophys. Biophys. Chem.*, **17**, 167–192

Watson, J. D. and Crick, F. H. C. (1953). A structure for deoxyribose nucleic acid. *Nature*, **171**, 737–738

Westhof, E., Romby, P., Romaniuk, P. J., Ebel, J., Ehresmann, C. and Ehresmann, B. (1989). Computer modeling from solution data of spinach chloroplast and of *Xenopus laevis* somatic and oocyte 5S rRNAs. *J. Mol. Biol.*, **207**, 417–431

Wittmann-Liebold, B. (1986). In Hardesty, B. and Kramer, G. (Eds.), *Structure, Function, and Genetics of Ribosomes*. Springer, New York, pp. 326–361

Woese, C. R., Gutell, R., Gupta, R. and Noller, H. F. (1983). Detailed analysis of the higher-order structure of 16S-like ribosomal ribonucleic acids. *Microbiol. Rev.*, **47**, 621–669

Wollenzien, P. L. and Cantor, C. R. (1982). Gel electrophoretic technique for separating crosslinked RNAs—application to improved electron microscopic analysis of psoralen crosslinked 16S ribosomal RNA. *J. Mol. Biol.*, **159**, 151–166

Wollenzien, P. L., Murphy, R. F., Cantor, C. R., Expert-Bezancon, A. and Hayes, D. H. (1985). Structure of the *Escherichia coli* 16S ribosomal RNA—psoralen crosslinks and *N*-acetyl-*N'*-(*p*-glyoxylylbenzoyl)cystamine crosslinks detected by electron microscopy. *J. Mol. Biol.*, **184**, 67–80

Woo, N. H., Roe, B. A. and Rich, A. (1980). Three-dimensional structure of *E. coli* initiator $tRNA^{fMet}$. *Nature*, **286**, 346–351

Wower, I. and Brimacombe, R. (1983). The localization of multiple sites on 16S RNA which are crosslinked to proteins S7 and S8 in *Escherichia coli* 30S ribosomal subunits by treatment with 2-iminothiolane. *Nucleic Acids Res.*, **11**, 1419–1437

Yonath, A., Leonard, K. R., Weinstein, S. and Wittmann, H. G. (1987). Approaches to the determination of the three-dimensional architecture of ribosomal particles. *Cold Spring Harbor Symp. Quant. Biol.*, **52**, 729–741

Zuker, M. (1989). On finding all suboptimal foldings of an RNA molecule. *Science*, **244**, 48–52

Zwieb, C. and Brimacombe, R. (1979). RNA–protein cross-linking in *Escherichia coli* 30S ribosomal subunits: precise localization of the nucleotide in 16S RNA which is coupled to protein S7 by ultraviolet irradiation. *Nucleic Acids Res.*, **6**, 1775–1790

5
Refinement of Three-dimensional Structures of Proteins and Nucleic Acids

Axel T. Brünger

INTRODUCTION

Molecular dynamics and Monte Carlo simulations can be helpful at stages of the experimental data analysis that are aimed at optimizing an atomic model which agrees with observed experimental information. One example is the refinement of three-dimensional structures of macromolecules based on X-ray crystallographic diffraction data (Brünger *et al.*, 1987b). Another example is the determination and refinement of three-dimensional structures of proteins and nucleic acids in solution by nuclear magnetic resonance spectroscopy (Wüthrich, 1989; Clore and Gronenborn, 1989). In both cases the problem can be formulated as to find the global minimum of a hybrid energy function

$$E_{\text{total}} = E_{\text{chem}} + wE_{\text{exp}} \tag{5.1}$$

where E_{chem} comprises empirical information about equilibrium covalent bonding geometry, vibrations, and hydrogen-bonded and non-bonded interactions in accordance with prior knowledge of these quantities obtained from small molecule crystallography and spectroscopy, and E_{exp} consists of difference expressions between observed and computed data which become zero if the atomic model matches the data. The weight w relates E_{exp} to E_{chem}. The notion of a hybrid energy function was first introduced by Jack and Levitt (1978) in the context of X-ray crystallographic refinement.

The global minimum of E_{total} is a three-dimensional structure of the atomic model which satisfies both the empirical as well as the experimental information. This is a highly non-linear optimization problem. In such a case, conventional gradient descent or least-squares optimization will fail unless the model is fairly close to optimal. Better convergence is achieved if, instead of restricting search directions to be energetically downhill, the optimization is allowed to go uphill as well, with probability $e^{-E/K_B T}$, where K_B is Boltzmann's constant and T is an effective temperature used as a control parameter. The optimization is controlled by the annealing schedule, defined as the sequence of temperatures and number of configurations sampled at each temperature during the search. This process, implemented as a Monte Carlo algorithm (Metropolis *et al.*, 1953), has been referred to as simulated annealing (SA) by Kirkpatrick *et al.* (1983). SA was first applied to combinatorial optimization problems, such as the 'travelling salesman' problem.

Monte Carlo and molecular dynamics simulations are both ways of generating conformations of the molecule that are consistent with a Boltzmann distribution appropriate to the specified temperature. For large biomolecular structures the molecular dynamics algorithm is generally more efficient at generating equilibrium structures (Northrup and McCammon, 1980). A molecular dynamics rather than a Monte Carlo algorithm was therefore used to refine protein structures with X-ray crystallographic diffraction data (Brünger *et al.*, 1987b). Similar techniques were used to determine and to refine three-dimensional structures of proteins in solution with interproton distance restraints derived from nuclear magnetic resonance (NMR) spectroscopy (Kaptein *et al.*, 1985; Clore *et al.*, 1985). We will refer to all these techniques simply as SA in spite of the fact that various other terms are used in the literature, such as restrained molecular dynamics (Kaptein *et al.*, 1985; Clore *et al.*, 1985) or dynamical simulated annealing (Clore and Gronenborn, 1989).

Introductions and reviews of general molecular dynamics techniques have been published elsewhere (McCammon and Harvey, 1987; Brooks *et al.*, 1988). In this review we shall focus on applications of SA, molecular dynamics, and energy minimization to specific non-linear optimization problems in structural molecular biology. First, we shall discuss general methodological aspects of the hybrid energy functions used in such non-linear optimization problems. We shall then present applications to the structure determination and refinement of small proteins and nucleic acids based on NMR interproton distance data. This is followed by applications to crystallographic refinement of macromolecules, which is aimed at obtaining the maximum information from the observed diffraction data and to evaluate its reliability. Finally, we present a generalization of the molecular replacement method to obtain initial phases for single-crystal diffraction data based on known structural information of homologous proteins.

GENERAL METHODOLOGY

Empirical Energy Function

The empirical potential energy E_{chem} is a function of all atomic positions of the system describing internal geometric interactions (bond lengths, bond angles, dihedral torsion angles, chiral centres, planarity of aromatic rings) as well as non-bonded (van der Waals, hydrogen bonding, and electrostatic) interactions:

$$E_{\text{chem}} = \sum_{\text{bonds}} k_{\text{b}}(r - r_0)^2 + \sum_{\text{angles}} k_\theta(\theta - \theta_0)^2$$

$$+ \sum_{\text{dihedrals}} k_\phi \cos(n\phi + d) + \sum_{\text{chiral, planar}} k_\omega(\omega - \omega_0)^2$$

$$+ \sum_{\text{atom pairs}} (ar^{-12} + br^{-6} + cr^{-1}) \qquad (5.2)$$

The parameters of the empirical potential energy E_{chem} are inferred from experimental as well as theoretical investigations (Lifson and Stern, 1982; Brooks *et al.*, 1983; Némethy *et al.*, 1983; Hermans *et al.*, 1984; Nilsson and Karplus, 1986; Weiner *et al.*, 1986). The SA program X-PLOR (Brünger, 1988a) by default uses the empirical energy parameters developed for the molecular dynamics program CHARMM, version 20 (Brooks *et al.*, 1983), although other energy functions, such as AMBER (Weiner *et al.*, 1986) can be emulated in X-PLOR. Unless otherwise noted, all applications discussed in this chapter made use of the polar-hydrogen and the all-hydrogen energy functions of CHARMM, version 20, for the crystallographic and the NMR spectroscopic applications respectively.

Modifications of the Empirical Energy Function

Modifications of the parameters of the empirical energy function are required to avoid unphysical conformation changes due to close contacts in the initial structure resulting from the effect of the experimental restraints (E_{exp}) or from the effects of high temperature during SA. For example, the force constants of the improper torsion angles that maintain handedness of chiral centres and the planarity of aromatic rings had to be increased by Brünger (1988b) to prevent distortions of these angles seen in X-ray crystallographic refinement when the original parameters were used. Similarly, the force constant specifying the dihedral torsion angle ω for the peptide bonds of all amino acids except proline was increased to prevent formation of *cis* peptide bonds. The force constant for the proline ω angle was kept at a small value to enable *cis–trans* transitions. However, experience has shown (Weis *et al.*, 1990) that this force constant should be increased during the final stages of refinement in order to obtain proper geometry around these peptide bonds. Similar adjustments of the empirical energy function are required for solution NMR structures (Nilges *et al.*, 1988b; Lee *et al.*, 1989) since the interproton distance restraints affect the geometry in the vicinity of hydrogen atoms that are involved in nuclear Overhauser enhancement (NOE) measurements.

Since bulk solvent is normally not part of the refinement of crystal structures or solution NMR structures, the charges of the headgroups of ionizable sidechains (Asp, Glu, Arg and Lys) have to be screened in order to avoid formation of artificial salt links with backbone atoms on the surface of the molecule. This can be accomplished by setting the charges to zero (Brünger, 1988b) or by neutralizing the charge (Fujinaga *et al.*, 1989).

In contrast to Fujinaga *et al.* (1989) we found it necessary to apply modifications of the empirical energy function even for room temperature calculations (300 K) or for straight minimizations. It can be argued that the use of stiffer parameters has a detrimental effect on the optimization by restricting the search to a smaller region of phase space. However, this is offset by high simulation temperatures and, furthermore, it is unlikely that regions with a large error in the initial model are improved by the more flexible parameters.

Concern has been raised about the introduction of bias into refined structures by using an empirical energy function as opposed to a 'geometric' energy function such as the one used in the restrained least-squares program PROLSQ (Konnert and Hendrickson, 1980). We define an energy function as geometric if it consists of terms for covalent bonds, bond angles, chirality, planarity and a repulsive non-bonded term, and if it does not contain electrostatic, dispersion or hydrogen bonding terms. The techniques of SA and the program X-PLOR (Brünger, 1988a) are not restricted to empirical energy functions. Nilges *et al.* (1988a) have developed a geometric energy function for X-PLOR in which identical energy constants were assigned to all covalent bonds (500 kcal mol^{-1} $Å^{-2}$), bond angles (500 kcal mol^{-1} rad^{-2}), and improper dihedral angles (500 kcal mol^{-1} rad^{-2}). The advantage of the repulsive non-bonded term is that it can be used with a small cutoff radius, thus reducing computing time, and that it allows atoms to pass through each other during the course of SA. The differences between a geometric energy function and an empirical energy function affect regions that are not well determined by the experimental information, such as disordered regions of crystal structures or surface loops in solution NMR structures. We observed little difference in well-defined regions. For instance, the root mean-square (RMS) difference for backbone atoms between a structure of crambin refined by

PROLSQ (Konnert and Hendrickson, 1980) and a structure refined by conjugate gradient minimization at 2 Å resolution using X-PLOR with the CHARMM polar-hydrogen empirical energy function was 0.05 Å (Brünger *et al.*, 1989a).

Gronenborn and Clore (1989) have discussed the role and the contribution of the empirical energy functions with respect to NMR-derived structures of oligonucleotides. They conclude that structural changes occurring during the course of SA are determined by the interproton distance restraints. The role of the energy function is to maintain near-perfect geometry and non-bonded interactions. However, features of the oligonucleotide that are less well determined by the interproton distance restraints, such as bending, can be affected by the empirical energy function.

Molecular Dynamics

Molecular dynamics simulations involve the simultaneous solution of the classical equations of motion for all atoms i of a macromolecule where the forces are derived from the total potential energy E_{total},

$$m_i \frac{\partial^2 r_i}{\partial t^2} = -\nabla_i E_{\text{total}} \tag{5.3}$$

The quantities r_i and m_i are the coordinates and masses respectively of atom i. $\nabla_i E_{\text{total}}$ can be computed analytically for all energy terms discussed in this review. The solution of these partial differential equations is normally carried out numerically by finite-difference methods, such as the Verlet algorithm (Verlet, 1967). The initial velocities are usually assigned to a Maxwellian distribution at the appropriate temperature.

Temperature Control

There are a number of ways of controlling the temperature during a molecular dynamics simulation. Three specific methods are discussed below: velocity scaling, Langevin dynamics and T-coupling.

Velocity Scaling

Velocity scaling consists of periodic uniform scaling of the velocities v_i, i.e.

$$v_i^{\text{new}} = v_i^{\text{old}} \sqrt{\frac{T}{T_{\text{curr}}}} \tag{5.4}$$

for all atoms i where T is the target temperature and T_{curr} is the current temperature. Each time velocities are scaled, the Verlet algorithm (Verlet, 1967) has to be restarted since it propagates positions without reference to velocities (Pastor *et al.*, 1988).

Langevin Dynamics
Langevin dynamics incorporates the influence of a heat bath into the classical equations of motion:

$$m_i \frac{\partial^2 r_i}{\partial t^2} = -\nabla_i E_{\text{total}} - m_i \gamma_i \frac{\partial r_i}{\partial t} + R(t) \tag{5.5}$$

where γ_i specifies the friction coefficient for atom i, and $R(t)$ is a random force. $R(t)$ is assumed to be uncorrelated with the positions and velocities of the atoms. It is described by a Gaussian distribution with mean of zero and variance

$$\langle R(t)R(t')\rangle = 2m_i\gamma_i K_B T \delta(t-t') \tag{5.6}$$

where K_B is Boltzmann's constant and $\delta(t-t')$ is the Dirac delta function. The finite-difference algorithm of Brünger *et al.* (1984) can be used to obtain a numerical solution of Equation (5.5) which is valid in the limit of $\gamma_i \Delta t \ll 1$. Pastor *et al.* (1988) showed that this algorithm is preferable compared with other finite-difference algorithms to propagate positions.

T-coupling
The weak temperature coupling method of Berendsen *et al.* (1984), which we will refer to as *T*-coupling, is related to Langevin dynamics except that it does not use random forces and it applies a global scale factor to the friction coefficient which is proportional to the difference between actual and target temperature:

$$m_i \frac{\partial^2 r_i}{\partial t^2} = -\nabla_i E_{\text{total}} - m_i \gamma_i v_i \left(1 - \frac{T}{T_{\text{curr}}}\right) \tag{5.7}$$

The second term on the right-hand side of Equation (5.7) is equivalent to a positive friction coefficient if $T_{\text{curr}} > T$, thus lowering the temperature; it produces a negative friction coefficient if $T_{\text{curr}} < T$, thus increasing the temperature. Equation (5.7) can be integrated numerically using the algorithm of Brünger *et al.* (1984).

Relative Weighting of E_{chem} and E_{exp}

The weight w (Equation 5.1) is needed in order to relate the effective energy term E_{exp} to the 'physical' energy term E_{chem}. The choice of w can be critical: if w is too large, the refined structure shows large deviations from ideal geometry; if w is too small, the refined structure does not satisfy the experimental information. Jack and Levitt (1978) proposed to set w so as to make the gradients of E_{chem} and E_{exp} of the same magnitude for the current structure. This approach implies that w has to be occasionally re-adjusted during the course of the refinement. We have succeeded in reducing the amount of re-adjustment necessary. Our method consists of a short molecular dynamics run with w set to zero starting with the initial structure, and then to adjust w as to make the magnitude of the gradients of E_{chem} and E_{exp} of the molecular dynamics structure equal (Brünger *et al.*, 1989a). The dynamics step is useful to randomize the structure so as to avoid the unrealistically low gradients that could occur if the initial structure were in a local minimum of E_{exp} or E_{chem}. This method has been used for SA refinements of crystal structures. However, at the final stages of refinement it may be necessary to fine tune w in order to optimize the geometry of the molecule. For example, in the case of the influenza virus haemagglutinin (Weis *et al.*, 1990) we have reduced the original weight by 10% to 40% during the final stages of refinement.

In the case of structure determinations and refinements of solution NMR structures, w is obtained by trial and error (Brünger *et al.*, 1986).

Classification for the Control Parameters of SA

The success of SA critically depends on the choice of the annealing schedule (Bounds, 1987), that is, the sequence of, and numerical values for, control parameters, such as temperature and relative weighting factors of E_{total}. We introduce in the following a classification scheme for the control parameters of SA that is specific for minimization problems involving a hybrid energy function E_{total}.

SA of Type 1

The basic idea of SA using molecular dynamics is to increase the kinetic energy of the atoms so as to ensure that the atoms can overcome local energy barriers of E_{total}. This approach was first applied to crystallographic refinement by Brünger *et al.* (1987b), and to NMR solution structure determination by Kaptein *et al.* (1985) and Clore *et al.* (1985). Equivalently, with $0 \leq c \leq 1$, E_{total} can be scaled by an overall factor c smaller than 1.0 while maintaining a constant temperature during the simulation:

$$m_i \frac{\partial^2 r_i}{\partial t^2} = -\nabla_i c E_{total} \tag{5.8}$$

$$E_{kin} = \sum_i \frac{1}{2} m_i \left(\frac{\partial r_i}{\partial t} \right)^2 = \text{constant} \tag{5.9}$$

This is equivalent to

$$m_i \frac{\partial^2 r_i}{\partial t'^2} = -\nabla_i E_{\text{total}} \tag{5.10}$$

$$E'_{\text{kin}} = \sum_i \frac{1}{2} m_i \left(\frac{\partial r_i}{\partial t'}\right)^2 = E_{\text{kin}}/c \tag{5.11}$$

with $t' = \sqrt{c}t$, i.e. the kinetic energy and, thus, the temperature are scaled by $1/c$. We will refer to procedures that scale E_{total} or, equivalently, the temperature T as SA of the first type.

SA of Type 2

In addition to the overall scaling of E_{total}, the weight w in front of E_{exp} (Equation 5.1) or weights of individual E_{exp} terms may be modified during the course of SA. In particular, w may be lowered while keeping the original E_{chem} term and maintaining a constant temperature at 300 K. We refer to weighting procedures involving E_{exp} alone as SA of the second type. This approach has been applied to the determination and refinement of solution NMR structures (Brünger *et al.*, 1986; Nilges *et al.*, 1988b), and to crystallographic refinement by Fujinaga *et al.* (1989). The approach is based on the assumption that it is the E_{exp} term that introduces large energy barriers which would normally require the use of very high temperatures, such as 1000–4000 K, to overcome them.

SA of Type 3

The third alternative is to control the weights for classes of partial energies of both E_{chem} and E_{exp} individually in order to ensure that the energy barriers for each class of partial energies are lower than the kinetic energy of the atoms. Examples of individually weighted terms are bond lengths, bond angles, diffraction data belonging to different resolution ranges, or short-range and long-range interproton distances. This third type of SA has been applied to solution NMR structure determination based on interproton distances starting from a random array of atomic coordinates (Nilges *et al.*, 1988a). It has not yet been applied to crystallographic refinement.

SOLUTION TO NMR STRUCTURE DETERMINATION AND REFINEMENT

The availability of high-field NMR spectrometers and the advent of two- and three-dimensional NMR spectroscopy have led to significant progress in the application of NMR to macromolecules (Ernst *et al.*, 1986; Oschkinat *et al.*, 1988). With the development of sequential resonance assignment strategies based on the delineation of through-bond and through-space (<5 Å) connectivities, it has become possible to obtain complete or virtually complete and unambiguous resonance assignments for small proteins up to 100 amino acids and oligonucleotides up to 16 base pairs and to derive a large set of approximate interproton distances by means of pre-steady-state NOE measurements (Wüthrich, 1986). These interproton distance estimates constitute the data for three-dimensional structure determination.

Since the available NOE distances (generally, <5 Å) do not fully determine the three-dimensional structure, the experimental information has to be augmented by information about the geometry and non-bonded interactions of the macromolecules. In other words, a hybrid energy function (Equation 5.1) is defined of which one tries to locate the global minimum. Several methods have been developed to locate a local minimum close to the global minimum; these methods have been reviewed by Braun (1987). In the following we will briefly describe the available methods and then focus on SA using molecular dynamics.

Methods

One class of methods operates in distance space and is generally referred to as metric matrix distance geometry (Crippen, 1977). The coordinates of the structure of n atoms are generated by a projection from $n(n-1)/2$-dimensional distance space into three-dimensional Cartesian coordinate space by solving a $n \times n$ dimensional eigenvalue problem. The mathematical formulation avoids all problems associated with 'folding' the macromolecules into the correct conformation. However, distance geometry is based on the assumption that all distances are known exactly and without error. This is never fulfilled for solution NMR structure determination because distances between heavy atoms are not observable, and only a subset of distances between hydrogen atoms are observable with limited precision. Nevertheless, the distance geometry method can be used to obtain an approximate solution to the non-linear optimization problem of Equation (5.1) when only observable interproton distances and geometric terms, such as bond lengths and bond angles, are taken into account. A C^{α}–C^{α} distance, for example, would initially be only restricted to an interval with the lower limit given by the sum of the van der Waals radii and the upper limit given by the maximum possible length of the polypeptide chain between the C^{α} atoms. Triangle inequalities are then used to restrict the distance intervals further, which is termed 'bound smoothing'. Even if all triangle inequalities are satisfied, the chosen distances still may not yield a consistent solution in three-dimensional Cartesian space. If they did, the eigenvalue problem of distance geometry would have exactly three non-zero solutions. For NMR applications there are many more non-zero solutions. Experience shows that one obtains a reasonable structure if one simply takes the three largest eigenvalues. The resulting three-dimensional structure will, however, violate many of the initial distance bounds. Another problem associated with distance geometry is that most algorithms only make use of information expressed as distances. A case in point is information about chirality. Distance geometry quite often produces mirror images (local or global) of the correct structure. These mirror images can only be rejected *a posteriori* by carrying out minimizations in Cartesian coordinates against E_{total} (Equation 5.1) which contains information about chirality. It is possible to use energy terms in distance space (Purisima and Scheraga, 1986), but this method is limited to small peptides since it is computationally very expensive.

Another class of methods operates in torsion angle space. The DISMAN method by Braun and Go (1985) minimizes Equation (5.1) in torsion angle space by employing the so-called variable target function approach. This method minimizes gradually a series of target functions which approximate Equation (5.1). For instance, for a polypeptide chain of n residues the target functions T_k ($k = 1, 2, \ldots, n$) only include distance information (NMR data or non-bonded interactions) for atom pairs belonging to residues with difference of their sequence number less than k. The strategy consists in first minimizing T_k with small values of k and then gradually increasing k up to n. The minimizations in DISMAN employ a conjugate gradient algorithm.

The ellipsoid algorithm (Billeter *et al.*, 1987) is a torsion angle minimization procedure that reduces the number of evaluations of E_{total} by treating the bounds on the values of selected interatomic distances as a convex programming problem. The method starts with an ellipsoid in torsion angle space such that the desired solution is certainly contained within it. Then, the volume of the ellipsoid is monotonically decreased by an iterative procedure. The centre of the ellipsoid will asymptotically reach a minimum. The ellipsoid method works best for low-dimensional problems, that is, 50 degrees of freedom or less (Billeter *et al.*, 1987).

Finally, the Monte Carlo algorithm has been applied to minimize E_{total} in torsion angle space (Bassolino *et al.*, 1988) starting from extended polypeptide chains. As was pointed out in the Introduction, the molecular dynamics algorithm is generally more efficient than the Monte Carlo algorithm for macromolecules at generating equilibrium structures (Northrup and McCammon, 1980). However, the Monte Carlo algorithm is advantageous for generating a large ensemble of different conformations of the molecule which is of importance when the accuracy of solution NMR structures has to be assessed.

A third class of methods operates in Cartesian coordinate space. The most prominent method here employs SA through molecular dynamics to refine protein (Kaptein *et al.*, 1985; Clore *et al.*, 1985) or nucleic acid (Nilsson *et al.*, 1986) structures or, more ambitiously, to determine protein structures starting from extended strands (Brünger *et al.*, 1986; Clore *et al.*, 1986a) or random atomic coordinates (Nilges *et al.*, 1988a). SA has also been combined with distance geometry calculations to combine efficiently the best features of both methods (Nilges *et al.*, 1988b). We shall focus in the following on SA through molecular dynamics and its applications.

Interproton Distance Energy Terms

Interproton distance restraints may be incorporated into Equation (5.1) as biharmonic effective potentials (Clore *et al.*, 1985; Nilsson *et al.*, 1986) when distances can be accurately determined (as in the case of DNA)

$$E_{\mathrm{exp}}^{\mathrm{NOE}}(r_{ij}) = \sum_{i,j} \begin{cases} u_{ij}(r_{ij} - r_{ij}^{\mathrm{obs}})^2 & \text{if } r_{ij} > r_{ij}^{\mathrm{obs}} \\ v_{ij}(r_{ij} - r_{ij}^{\mathrm{obs}})^2 & \text{if } r_{ij} \leq r_{ij}^{\mathrm{obs}} \end{cases} \tag{5.12}$$

where the sum proceeds over all pairs i,j of protons between which NOEs are observed; r_{ij} and r_{ij}^{obs} are the calculated and observed interproton distances respectively. The constants u_{ij}, v_{ij} are given by

$$u_{ij} = k_{\mathrm{B}}T\frac{1}{(2\Delta_{ij}^{+})^2} \tag{5.13}$$

$$v_{ij} = k_{\mathrm{B}}T\frac{1}{(2\Delta_{ij}^{-})^2} \tag{5.14}$$

where k_{B} is the Boltzmann constant, T is the temperature of the simulation, and Δ_{ij}^{+} and Δ_{ij}^{-} are the positive and negative error estimates respectively of r_{ij}. If only ranges of distances can be specified as in the case of proteins, the interproton distances can be incorporated in the form of double-well potentials (Clore *et al.*, 1986b)

$$E_{\mathrm{exp}}^{\mathrm{NOE}}(r_{ij}) = \sum_{i,j} \begin{cases} c_{ij}(r_{ij} - r_{ij}^{\mathrm{u}})^2 & \text{if } r_{ij} > r_{ij}^{\mathrm{u}} \\ 0 & \text{if } r_{ij}^{\mathrm{l}} \leq r_{ij} \leq r_{ij}^{\mathrm{u}} \\ c_{ij}(r_{ij} - r_{ij}^{\mathrm{l}})^2 & \text{if } r_{ij} < r_{ij}^{\mathrm{l}} \end{cases} \tag{5.15}$$

where the sum proceeds over all pairs i,j of protons between which NOEs are observed; r_{ij}^{u} and r_{ij}^{l} are the values of the upper and lower limits of the target distances respectively and c_{ij} are individual weights. Torsion angle restraints derived from coupling constants measurements using the Karplus relationship (Karplus, 1959) can also be incorporated in the form of double-well potentials (Clore *et al.*, 1986b):

$$E_{\text{exp}}^{\text{coupling}}(r_{ij}) = \sum_i \begin{cases} k_i(\phi_i - \phi_i^{\text{u}})^2 & \text{if } \phi_i > \phi_i^{\text{u}} \\ 0 & \text{if } \phi_i^{\text{l}} \leq \phi_i \leq \phi_i^{\text{u}} \\ k_i(\phi_i - \phi_i^{\text{l}})^2 & \text{if } \phi_i < \phi_i^{\text{l}} \end{cases} \tag{5.16}$$

where the sum proceeds over all torsion angles i for which coupling constants have been determined; ϕ_i^{u} and ϕ_i^{l} are the upper and lower limits of the target range of a particular torsion angle, and ϕ_i is its calculated value.

For distances involving either methylene protons without stereospecific assignments or methyl protons, one can argue that single $\langle r_{ij}^{-6}\rangle^{-1/6}$ mean distances should be used as this is the quantity that can be related directly to the experimental NOE. This quantity is heavily weighted towards the distance with the smaller value and thus may lock the structure into an incorrect conformation. An alternative approach is to compute geometric centre-averaged distances between groups of protons and to increase the allowed distance range for the NOEs between groups of protons. This approach, which is equivalent to the pseudo-atom method (Wüthrich, 1986), has the advantage of avoiding local minima for alternative conformations of groups involving protons without stereospecific assignments, but clearly some information is lost. Recently, stereospecific assignments of methylene groups have become feasible in a few cases (Hyberts *et al.*, 1987) in which case no averaging is required.

Annealing Schedules

Refinement of pre-folded structures obtained by distance geometry calculations is normally carried out at a low temperature (300–1000 K). The weights for all terms of E_{total} are kept constant during the refinement, i.e. SA of the first type is used. If partial or complete folding or re-folding of the structure is required, one can either reduce the weights for individual energy terms with large initial gradients, or increase the simulation temperature, or both. The first such published SA protocol was of the second type: it used a two-step procedure where during the first step only secondary structure interproton distance restraints were included and in the second step the remaining interproton distance restraints were included while the weights for all E_{chem} terms were kept constant (Brünger *et al.*, 1986). This protocol had a relatively small success rate ($\approx$50%). A number of new protocols were developed to improve the success rate (Clore and

Gronenborn, 1989). The most promising approach was taken by Nilges *et al.* (1988a) where the weights for all partial E_{exp} and E_{chem} terms (such as bond lengths, bond angles, classes of interproton distances) were controlled individually in order to ensure that the energy barriers for each class of partial energies are lower than the kinetic energy of the atoms (SA of the third type).

Applications

The radius of convergence of SA using molecular dynamics is quite large. Figure 5.1 shows the pathway of SA for crambin starting from an extended

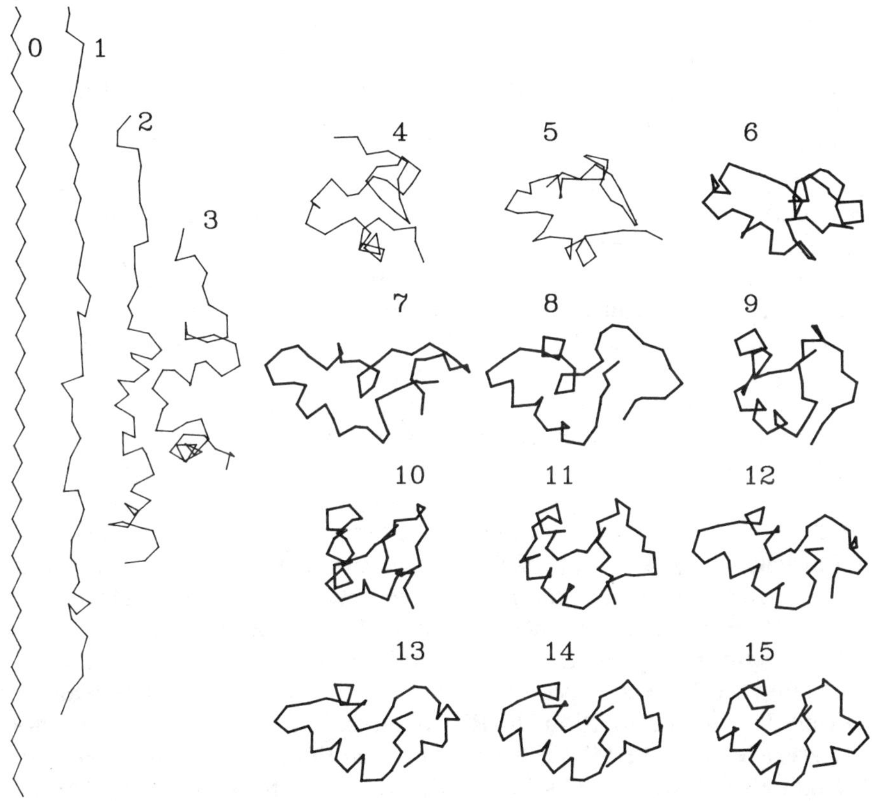

Figure 5.1 Pathway of SA for the structure determination of crambin starting with an extended β-strand using interproton distance data (Brünger *et al.*, 1986). Simulated interproton distance data were used that could realistically be obtained by two-dimensional NOE spectroscopy. Snapshots are shown at 1 ps intervals; only the C^{α} backbone is shown. During the first 5 ps (thin lines) only interproton distance restraints between residues i, $i \pm 1$, $i \pm 2$, $i \pm 3$, $i \pm 4$, $i \pm 5$ were taken into account, while all interproton distances were used during the subsequent stages (thick lines)

β-strand (Brünger *et al.*, 1986). The conformational changes were very rapid. At the end of the first 2 ps of SA, the secondary folding was essentially established while the molecule was still in an extended conformation; at 9 ps, the tertiary structure was approximately correct. Convergence to the final structure was achieved after 15 ps.

Another example of the radius of convergence of SA was provided in the case of nucleic acids by Nilsson *et al.* (1986). Convergence to a B-DNA conformation was achieved starting from an A-DNA conformation and using idealized distances for B-DNA, or when going from B-DNA to A-DNA using A-DNA distances. Similar convergence has been achieved in structure determinations of various oligonucleotides (up to dodecamers) using solution NMR data starting from A-DNA and B-DNA conformations (Clore and Gronenborn, 1989). In addition to interproton distance restraints, torsion angle restraints for the C4′–C3′ (δ) bonds derived from coupling constant measurements can be necessary to avoid formation of local mirror images of the backbone conformation (Clore *et al.*, 1988).

The ultimate convergence test of SA was performed by starting from a random array of atomic coordinates (Nilges *et al.*, 1988a) using an SA protocol of the third type. Convergence starting from random coordinates was achieved in almost all cases studied. Furthermore, it was possible to correct a misfolded structure of crambin which consisted of a mirror image of the crystal structure. This mirror image represented a deep local minimum that had caused previous protocols to fail (Nilges *et al.*, 1989).

Because of the lack of a simple theory that relates the observed NMR spectral intensities and the atomic coordinates of an atomic model, the accuracy of NMR structure determination is at present assessed by generating a family of conformers each of which satisfies the experimental as well as the chemical information using different refinement protocols or starting conditions. It is argued that the smaller the spread of the conformers, the more accurately the structure is determined (Havel and Wüthrich, 1985). This approach assumes that the sampling of conformational space is adequate, that is a representative sample of structures have to be obtained that satisfy the experimental information.

The 74 residue protein α-amylase inhibitor tendamistat (Kline *et al.*, 1988) and the 43 residue protein BDS-I (Driscoll *et al.*, 1989) are examples of well-determined solution NMR structures. For the α-amylase inhibitor, the average atomic RMS difference between the individual structures and the mean structure is 0.57Å for backbone atoms and 1.01 Å for all atoms. For BDS-I, the corresponding values are 0.67 Å and 0.90 Å. In marked contrast, the 1–29 fragment of human growth hormone releasing factor in trifluoroethanol shows RMS differences for backbone atoms of up to 6 Å between individual conformers (Brünger *et al.*, 1987a). A series of three-dimensional structures were determined by SA and distance geometry (Figure 5.2). The structures satisfy the NOE distance data, and have

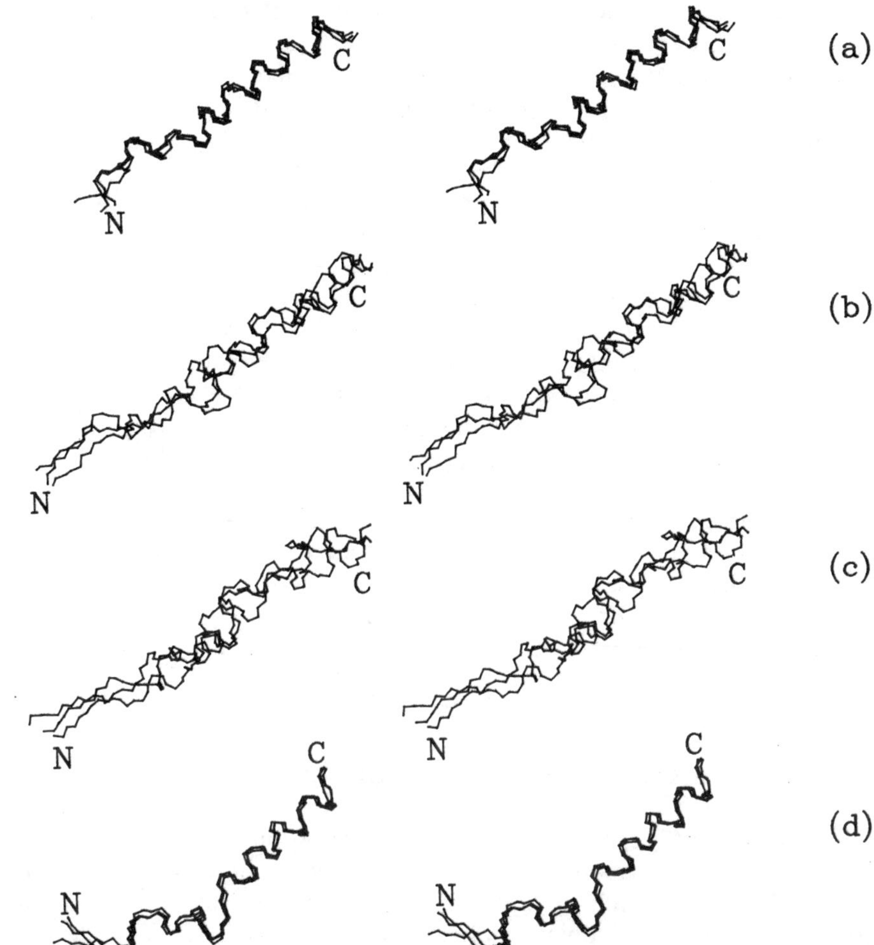

Figure 5.2 Stereo pictures showing best fit superpositions of solution conformations using NOE data (Brünger *et al.*, 1987a) of the 1–29 fragment of human growth hormone releasing factor. Only the backbone (C, C^{α}, N) is shown: (a) SA structures starting from an extended α-helix; (b) SA structures starting from an extended β-strand; (c) SA structures starting from a polyproline helix; (d) SA structures starting from a mixed α–β structure; (e) distance geometry structures using DISGEO (Havel, 1986)

reasonable geometries and empirical energies. Because no long-range NOEs were observed, the NOE data were not sufficient for global convergence to a unique three-dimensional geometry. The reason for the absence of observable tertiary NOEs could be one of two indistinguishable possibilities: the structures are extended or the NOE data set represents an average over a set of distinct folded conformations.

SA starting from several different initial structures appears better suited than distance geometry for exploring possible conformations in cases where NMR information is incomplete. The DISGEO (Havel, 1986) structures were biased towards structures that were obtained by SA when starting from an extended β-strand (Figure 5.2). SA with other initial conditions (e.g. α-helix) produced structures that were not obtained by the DISGEO calculations. From the DISGEO calculations alone one might have concluded that there is a unique conformation of the molecule. Analysis of the DISGEO sampling problem revealed that, during the first stage of the distance geometry calculation, distances were not chosen uniformly from the initial distance ranges but rather a Gaussian distribution was assumed around the distance centroid (M. Nilges, personal communication). This feature, which can easily be removed from the program, improves the convergence properties of DISGEO for globular proteins but it introduces a bias for extended structures.

CRYSTALLOGRAPHIC REFINEMENT OF MACROMOLECULES

The goal of crystallographic structure determination is to obtain a three-dimensional atomic model of the macromolecule that agrees with the observed diffraction information and that is chemically reasonable, i.e. the goal is to find a global minimum of Equation (5.1). The atomic models are derived from electron density maps that are computed using the observed structure factor amplitudes of the crystal and phase information obtained, for instance, through multiple isomorphous replacement (MIR) (Watenpaugh, 1985). The accuracy of the initial atomic model is usually limited, partly because the phase information is insufficient to show atoms individually and partly as a result of errors during the initial model building. This limited accuracy affects the understanding of the chemistry of the macromolecule and it is therefore important to improve the atomic models of macromolecules by crystallographic refinement.

Crystallographic refinement of macromolecules involves a series of steps, each of which consists of a round of minimization of E_{total} (Equation 5.1) followed by re-fitting the model structure to difference electron density maps with interactive computer graphics (Jones, 1978). During the final stages of refinement, solvent molecules are usually included and alternative conformations for some atoms or residues in the protein may be introduced. With currently available computing power, manual adjustments using computer graphics to display and move positions of atoms of the model in the electron density maps constitute the rate-limiting step in the refinement process and, thus, the need for automation or at least reduction of the manual steps arises.

We will first briefly describe various refinement techniques to minimize E_{total} and then focus on SA refinement.

Methods

Several algorithms have been developed over the past 20 years to minimize E_{total} (Stout and Jensen, 1989). These algorithms can be generally classified into constrained or restrained least-squares minimization, conjugate gradient minimization, and SA refinement. Algorithms for the refinement of individual atomic parameters as well as of rotational and translational degrees of freedom for rigid groups of atoms have been developed.

The stereochemical restraint refinement program PROLSQ (Konnert and Hendrickson, 1980) minimizes E_{total} by non-linear least-squares minimization. All chemical information in E_{chem} is expressed in terms of squared differences between idealized and calculated geometric quantities, such as bond lengths, bond angles, chiral volumes and out-of-plane displacements. In other words, PROLSQ uses a geometric energy function (see General Methodology section). Application of full-matrix least-squares minimization (Press *et al.*, 1986) is computationally expensive. Therefore, only diagonal terms and off-diagonal terms involving chemical restraints are taken into account. This sparse least-squares matrix is inverted by a conjugate gradient minimization procedure. This must not be confused with conjugate gradient minimization of E_{total}.

The constrained–restrained program CORELS (Sussman *et al.*, 1977) is closely related to PROLSQ except that groups of atoms can be defined as rigid bodies, and bond lengths and angles can be held fixed to reduce the number of degrees of feedom for low-resolution applications. The program TNT (Tronrud *et al.*, 1987) employs a conjugate gradient minimization procedure to minimize E_{total}. The program EREF (Jack and Levitt, 1978) utilizes an empirical energy function for E_{chem} and employs conjugate gradient minimization to minimize E_{total}. Finally, crystallographic SA refinement (program X-PLOR; Brünger, 1988a) consists of minimizing E_{total} using molecular dynamics (Brünger *et al.*, 1987b). We shall focus on this method in the following.

Crystallographic Energy Terms

Crystallographic Residual

The crystallographic residual is defined as the sum over the squared differences between the observed ($|F_{obs}(h, k, l)|$) and calculated ($|F_{calc}(h, k, l)|$) structure factor amplitudes

$$E^{A}_{exp} = \sum_{h,k,l} [|F_{obs}(h, k, l)| - k|F_{calc}(h, k, l)|]^2 \qquad (5.17)$$

where h, k, l are the reciprocal lattice points of the crystal, and k is a scale factor. In restrained least-squares refinement k is treated as an independently minimized parameter whereas in SA refinement it is set to the value which makes the derivative of E^{A}_{exp} with respect to k zero (Brünger *et al.*, 1989a):

$$k = \sum_{\boldsymbol{h}} W_{\boldsymbol{h}}|F_{obs}(\boldsymbol{h})||F_{calc}(\boldsymbol{h})| \,/\, \sum_{\boldsymbol{h}} W_{\boldsymbol{h}}F_{calc}(\boldsymbol{h})^2 \qquad (5.18)$$

This is a necessary condition to make E^{A}_{exp} minimal. If k were not to be eliminated from Equation (5.17) it would be necessary to optimize k periodically during the molecular dynamics simulation.

The structure factors ($F_{calc}(\boldsymbol{h})$) of the atomic model can be expressed in the following way:

$$F_{calc}(\boldsymbol{h}) = \sum_{s} \sum_{i} q_i f_i(\boldsymbol{h}) \exp[-B_i(\mathscr{F}^* \cdot \boldsymbol{h})^2/4] \exp[2\pi i \boldsymbol{h} \cdot (\mathbb{O}_s \cdot \mathscr{F} \cdot \boldsymbol{r}_i + \boldsymbol{t}_s)] \qquad (5.19)$$

The first sum extends over all symmetry operators ($\mathbb{O}_s$, $\boldsymbol{t}_s$) composed of the matrix $\mathbb{O}_s$ representing a rotation and a vector $\boldsymbol{t}_s$ representing a translation. The second sum extends over all unique atoms i of the system. The quantity $\boldsymbol{r}_i$ denotes the orthogonal coordinates of atom i in Å. $\mathscr{F}$ is the 3×3 matrix that converts orthogonal coordinates into fractional coordinates; $\mathscr{F}^*$ denotes the transpose of it. The columns of $\mathscr{F}^*$ are equal to the reciprocal unit cell vectors $\boldsymbol{a}^*$, $\boldsymbol{b}^*$, $\boldsymbol{c}^*$. B_i is the atomic temperature factor and q_i is the occupancy for atom i. The atomic form factors $f_i(\boldsymbol{h})$ are approximated by an expression consisting of four Gaussians and a constant:

$$f_i(\boldsymbol{h}) = \sum_{k=1}^{4} a_{ki} \exp[-b_{ki}(\mathscr{F}^* \cdot \boldsymbol{h})^2/4] + a_{0i} \qquad (5.20)$$

The constants a_{ki} and b_{ki} can be obtained from the *International Tables for X-ray Crystallography* (1974).

Equation (5.19) represents a simple way to compute the structure factors and their derivatives of the atomic model, but it is computationally expensive. In principle, the structure factors and their derivatives would have to be recomputed at each time step of the numerical integration of Equation (5.3). To reduce the computational requirements of SA refinement two approximations were introduced by Brünger *et al.* (1989a). The first approximation consists of keeping the first derivatives of E^{A}_{exp} constant until any atom has moved by more than Δ_F relative to the position at which the derivatives were last computed. At that point all derivatives are updated. Tests have shown that for Δ_F less than 0.3 Å this approximation does not affect the success of SA refinement and it speeds up the computation by almost an order of magnitude. For minimizations Δ_F has to be set to very small values in order to ensure that the gradient of E^{A}_{exp} is a smooth function of the displacement vector used in the minimizing algorithm. This is particularly important during the final cycles of the minimization.

The second approximation consists of computing $F_{calc}(\boldsymbol{h})$ by numerical evaluation of the atomic electron density on a finite grid followed by fast-Fourier transformation (FFT). The FFT method provides a way to speed up the calculation by at least an order of magnitude (Ten Eyck, 1973; Brünger, 1989).

Phase Restraints

A phase restraint term in crystallographic refinement is related to the technique of phase combination (Stuart and Artymiuk, 1985) which is sometimes used to improve the quality of the initial electron density maps. Since the experimental phase information for macromolecules is usually not very accurate, the errors of the phase observations have to be taken into account. Ideally, this would be accomplished by computing a phase probability distribution for each reflection and to refine the model phases against these distributions. A general representation of phase probability distributions is discussed by Hendrickson and Lattman (1970). Rees and Lewis (1983) assumed a simple harmonic distribution where the width of the potential was determined by the figure of merit. This choice of potential suffers from a bias towards the phase centroid. A simple and yet bias-free representation consists of a square-well potential around the phase centroid where the width of the square-well is determined by the individual figure of merit:

$$E^{P}_{exp} = W_P/N_P \sum_{\boldsymbol{h}} W_{\boldsymbol{h}}\, S\{\text{modulo}_{2\pi}[\phi_{obs}(\boldsymbol{h}) - \phi_{calc}(\boldsymbol{h})],\ \arccos[m(\boldsymbol{h})]\} \quad (5.21)$$

The sum extends over all observed reflections with indices $\boldsymbol{h} = (h, k, l)$ for which phase information is available. W_P is an overall weight factor, N_P is a normalization factor set equal to the number of phase specifications occurring in the sum, $W_{\boldsymbol{h}}$ are individual weights for each reflection $\boldsymbol{h}$, $\phi_{obs}(\boldsymbol{h})$ is the phase centroid obtained from MIR or other methods,

$\phi_{\text{calc}}(\boldsymbol{h})$ is the phase of the calculated structure factors, $m(\boldsymbol{h})$ is the individual figure of merit, and $S\{x,y\}$ is a double-well function

$$S\{x,y\} = \begin{cases} (x-y)^2 & x>y \\ 0 & -y<x<y \\ (y+x)^2 & -y>x \end{cases} \tag{5.22}$$

This form of the effective potential energy $E^{\text{P}}_{\text{exp}}$ ensures that the calculated phases are restrained only within the experimental error limits given by the figure of merit for each reflection, i.e. as long as a calculated phase $\phi_{\text{calc}}(\boldsymbol{h})$ is within the range $\phi_{\text{obs}}(\boldsymbol{h}) \pm \arccos[m(\boldsymbol{h})]$ no restraint is applied to that particular phase. It should be noted that Equation (5.21) is not restricted to MIR phases but can be used with any phase information.

Another approach is taken by Arnold and Rossmann (1988), who do not separate the amplitude and phase information, but rather restrain the real and imaginary parts A,B of the structure factor simultaneously. This has the advantage that the ambiguity in the direction of the phase shift when the difference is close to 180° is resolved. The simultaneous A,B restraints could be incorporated straightforwardly into SA refinement since the SGFFT algorithm (Brünger, 1989) is independent of the choice of the target function.

Crystallographic Symmetry Interactions

Experience has shown that the influence of crystal packing interactions can be important in crystallographic refinement. In the absence of crystal packing information, backbone or sidechain atoms of the primary molecule can penetrate symmetry-related molecules. This can occur in contact regions between molecules where the electron density does not clearly define the molecular boundaries. An energy term which describes non-bonding interactions between the molecule(s) located in the asymmetric unit and all crystallographic symmetry-related molecules surrounding the asymmetric unit was derived by Brünger *et al.* (1989a):

$$E^{\text{crystal}}_{\text{chem}} = \sum_{s} \sum_{i \geq j} \text{NB}\{|\mathcal{F}^{-1} \cdot \text{MinG}(\mathcal{F} \cdot \boldsymbol{r}_i - \mathcal{O}_s \cdot \mathcal{F} \cdot \boldsymbol{r}_j + \boldsymbol{t}_s)|\} \tag{5.23}$$

where $\mathcal{F}$ is the matrix that converts orthogonal coordinates into fractional coordinates, the first sum extends over all symmetry operators $(\mathcal{O}_s, \boldsymbol{t}_s)$ of the crystal, and the second sum extends over all pairs of atoms (i,j) for which the argument of the function NB { } is less than a specified cutoff r_{cut}, and $\boldsymbol{r}_i$ and $\boldsymbol{r}_j$ are the coordinates of atoms i and j respectively. The function $\text{MinG}(\boldsymbol{r})$ defines the minimum image distance in fractional coordinate space. It operates on each component of the three-dimensional vector $\boldsymbol{r}$ separately where the operation on each component x is given by

$$\text{MinG}(x) = \text{sign}(-x)\,\text{int}(|x| + 1/2) + x \qquad (5.24)$$

The function int(y) is defined as the integer part of y, and sign(y) is defined as the sign of y. For the application of Equations (5.23) and (5.24) to compute the crystal symmetry interactions, one has to make the reasonable assumption that the cutoff r_{cut} is less than half of any of the edges of a maximal orthogonal box fitting into the unit cell of the crystal. In the special case of an orthorhombic space group r_{cut} has to be less than half of any of the unit cell vectors of the crystal. NB$\{r\}$ is a non-bonding interaction potential, i.e. it is a sum of van der Waals and electrostatic interactions:

$$\text{NB}\{r\} = ar^{-12} + br^{-6} + cr^{-1} \qquad (5.25)$$

where the functional form of NB$\{r\}$ is chosen to be indentical to the potential used for intramolecular interactions (Equation 5.2). Thus, the non-bonding interaction energy between two atoms is independent of whether it is an intermolecular or an intramolecular interaction.

Non-crystallographic Symmetry Constraints–Restraints

The existence of non-crystallographic symmetry (NCS) in a crystal can be exploited to average noise in the data and, thus, to obtain structures that are better defined from what can be expected at the particular resolution the crystal diffracts to. A case in point are virus structures where the fact that the problem is overdetermined through NCS can be utilized to assist during the primary phasing (Champness *et al.*, 1976; Rossmann *et al.*, 1985).

NCS can be utilized in SA refinement by two different methods, which we termed 'strict' and 'restrained' (Weis *et al.*, 1990). Strict NCS assumes that the monomers (or, more generally, protomers) are strictly identical, thereby allowing SA refinement of the protomer only. The NCS operators (rotation matrices and translation vectors) that generate the crystallographic asymmetric unit from the protomer coordinates are applied to the protomer coordinates to generate the structure factors and their first derivatives with respect to the atomic parameters for the entire crystallographic asymmetric unit. The strict NCS method thus improves the signal-to-noise ratio by averaging the noise in the data. For non-bonded interactions, the NCS operators and additionally, in the general case, a subset of the crystallographic symmetry operators needed to define completely the internal symmetry of the oligomer can also be used to compute van der Waals and electrostatic interactions between the protomer and the atoms of the NCS-related atoms. Intersubunit interactions are thus correctly treated, while crystal symmetry non-bonded interactions cannot be treated as atoms in lattice contacts are in non-equivalent chemical environments with respect to their NCS mates.

Restrained NCS has been incorporated into SA refinement similar to the implementation of NCS in restrained least-squares refinement by Hendrickson (1985). The entire crystallographic asymmetric unit is refined, with NCS-related atoms restrained to their average positions $\bar{r}_i = (\bar{x}, \bar{y}, \bar{z})$ after least-squares superposition of the protomers by adding an effective energy term to Equation (5.1):

$$E_{\mathrm{chem}}^{\mathrm{NCS}} = k_{\mathrm{NCS}} \sum_{n \in \mathrm{NCS}} \sum_{i} (r_{i,n} - \bar{r}_i)^2 \qquad (5.26)$$

The first sum extends over all NCS-related molecules n while the second sum extends over all unique atoms i of the protomer. Restrained NCS is useful when one wishes to impose NCS on only part of the structure or to restrain parts of the structure more tightly than others.

Annealing Schedules

The influence of the temperature control method, weighting of the energy terms, cooling rate, and duration of the heating stage on the success of SA refinement has been studied by Brünger *et al.* (1990). It was concluded that *T*-coupling is preferable to velocity scaling since the latter sometimes causes large temperature fluctuations during molecular dynamics. *T*-coupling is also preferable to Langevin dynamics in the context of SA since the friction term of the latter slows the atomic motions. Slow-cooling protocols (typically 25 K decrements every 25 fs) produce lower *R* factors than faster-cooling protocols. The first type of SA produces somewhat lower *R* factors than SA of the second type. If the cooling is sufficiently slow, the initial heating stage is not required. Repeating the slow-cooling process improves the results somewhat. Based on these results the slow-cooling protocol described in Table 5.1 was recommended.

Table 5.1 Recommended slow-cooling protocol

Stage	*Description*
1	Determination of weights
2	Minimization, 40 conjugate gradient steps with soft repulsive potential followed by 120 conjugate gradient steps with van der Waals non-bonded potential, C^{α} restraints at 20 kcal mol^{-1} Å^{-2}, $\Delta_F = 0.05$ Å
3	Slow-cooling molecular dynamics, 3.7 ps, $T = 4000$–300 K (or lower initial temperature if instabilities occur), time step = 0.5 fs, *t*-coupling, friction coefficient $\gamma_i = 100$ ps^{-1}, temperature decrement 25 K every 25 fs, $\Delta_F = 0.2$ Å
4	Minimization, 80 conjugate gradient steps, $\Delta_F = 0.0$ Å

Applications

A number of tests on crambin (Brünger *et al.*, 1989a), aspartate aminotransferase (Brünger, 1988b), myohemerythrin (Kuriyan *et al.*, 1989), and phospholipase A2 (Fujinaga *et al.*, 1989) have shown that crystallographic SA refinement starting from initial models obtained by MIR or molecular replacement produces a structure that has an overall *R* factor and geometry significantly improved compared with conventional procedures without model building. SA refinement has already been used to refine a number of new structures, including foot-and-mouth disease virus (Aharya *et al.*, 1989), the human immunodeficiency virus encoded protease (Navia *et al.*, 1989; Wlodawer *et al.*, 1989), the membrane-pore-forming fragment of colicin A (Parker *et al.*, 1989), tumour necrosis factor (Jones *et al.*, 1989), human renin (Sielecki *et al.*, 1989), streptavidin (Weber *et al.*, 1989), influenza virus haemagglutinin (Weis *et al.*, 1990), aspartate carbamoyltransferase (Gouaux and Lipscomb, 1988), and the nucleic acid dodecamer CGCATATATGCG (Yoon *et al.*, 1988).

SA makes crystallographic refinement more efficient by moving sidechain atoms by more than 2 Å, changing backbone conformations, or flipping peptide bonds. Figure 5.3 shows three representative cases where SA refinement has essentially converged to a manually refined structure of the enzyme aspartate aminotransferase. The Pro-138 residue was initially built in the *trans* conformation (Figure 5.3(a)). Restrained least-squares refinement without re-fitting kept the proline residue in the *trans* conformation. SA refinement changed the conformation to *cis*. This was also found independently through manual re-fitting. It should be noted that Pro-138 has been found to be in the *cis* conformation in all X-ray structures of aspartate aminotransferases that have been determined to date. In the case of His-193 (Figure 5.3(b)), the imidazole ring has undergone a 90° rotation around the χ_1 bond during SA refinement. This rotation was accompanied by significant structural changes of the backbone atoms. This resulted in convergence of the SA-refined structure to the manually refined structure. In the case of Glu-265 (Figure 5.3(c)), SA refinement has flipped the peptide bond by 180°, similar to the result obtained by manual re-fitting. In both cases, the structural changes were not accomplished by restrained least-squares refinement without re-fitting.

SA refinement is not able to correct large erroneous chain traces. The backbone conformation of residues Phe-223 to Ala-229 of the SA-refined structure (Figure 5.4(c)) has not converged to that of the manually refined structure (Figure 5.4(d)). The chain trace of the initial structure contained a gross error which interchanged residues Gln-226 and Phe-228, presumably because of the poor quality of the MIR map (Figure 5.4(a)). The interchange of the two residues is too large a motion to obtain from a short 3.0 ps molecular dynamics simulation. However, the electron density map

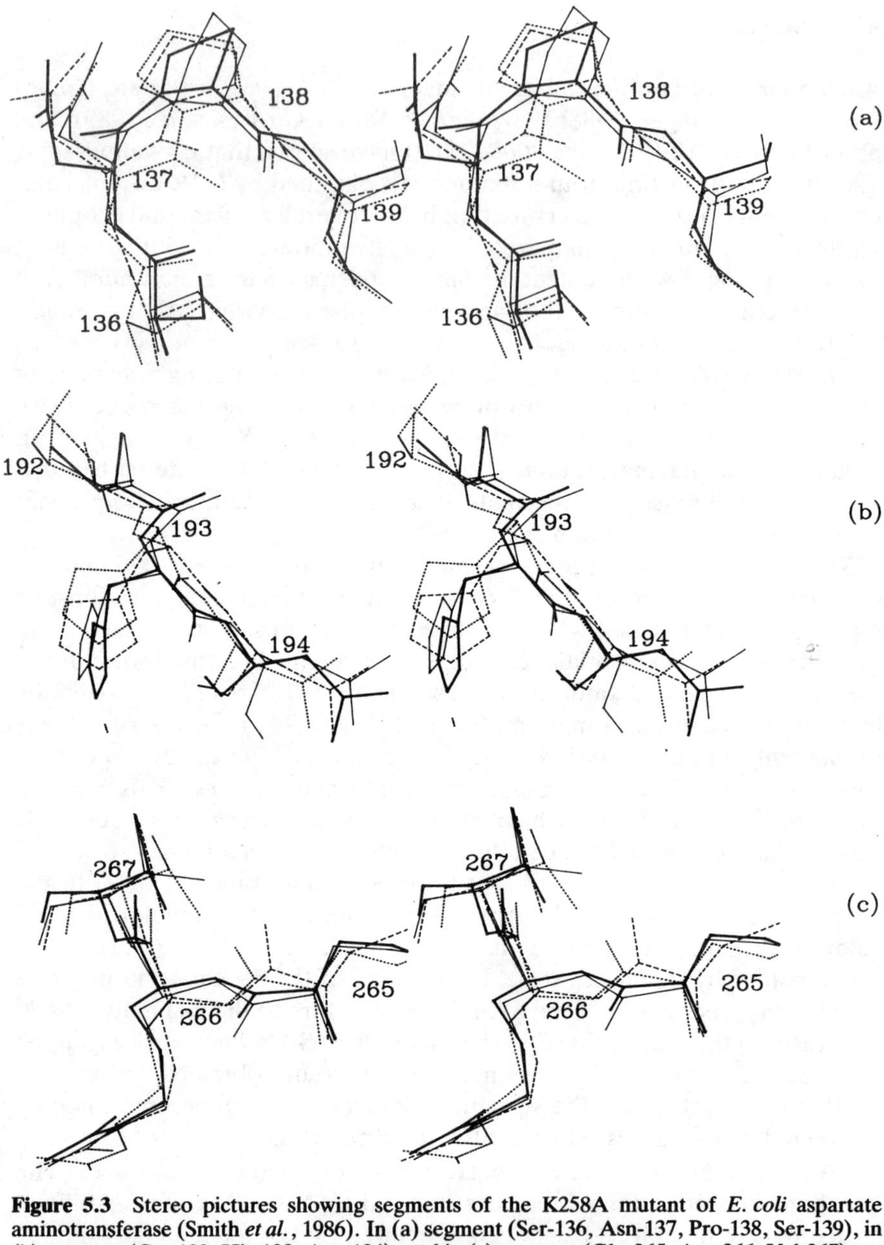

Figure 5.3 Stereo pictures showing segments of the K258A mutant of *E. coli* aspartate aminotransferase (Smith *et al.*, 1986). In (a) segment (Ser-136, Asn-137, Pro-138, Ser-139), in (b) segment (Cys-192, His-193, Asn-194), and in (c) segment (Glu-265, Arg-266, Val-267) are shown. Superimposed are a manually refined structure (thick lines), the SA-refined structure (thin lines), the restrained least-squares refined structure (broken lines), and the initial structure (dotted lines). All refinements (Brünger, 1988b) started from the same initial structure obtained by MIR

(a)

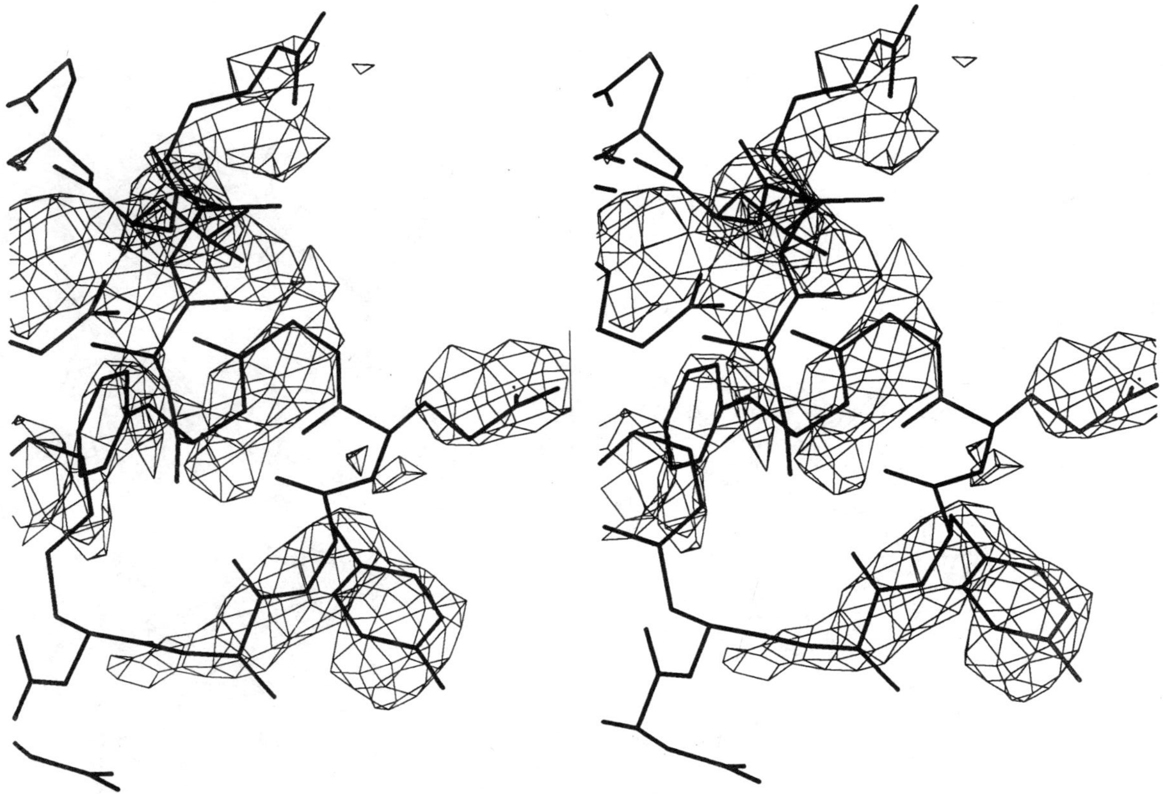

Figure 5.4 Stereo pictures showing a segment of aspartate aminotransferase, which consists of residues 223–229 (Phe-223, Ala-224, Tyr-225, Gln-226, Gly-227, Phe-228, Ala-229). In (a) the initial structure, in (b) the restrained least-squares refined structure, in (c) the SA-refined structure, and in (d) the manually refined structure are shown. The atomic models are shown as thick lines. Electron density maps are superimposed as thin lines. In the case of (a) the map was computed from F_{obs} amplitudes and MIR phases at 3 Å resolution. In the case of (b), (c) and (d) the maps were computed from $2F_{obs} - F_{calc}$ amplitudes using F_{calc} phases corresponding to the atomic models at 2.8 Å resolution. The maps were contoured at 1.5σ where σ is the RMS electron density in the unit cell

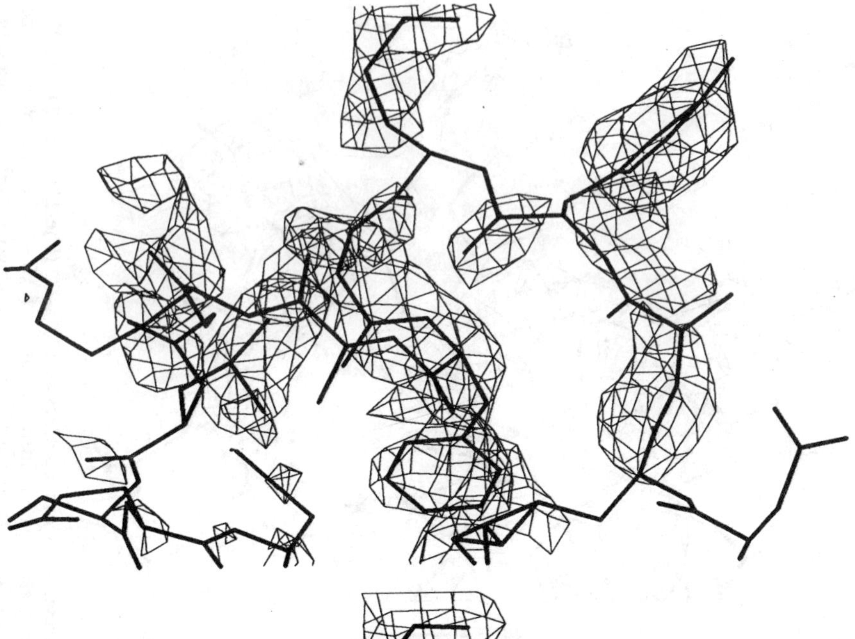

(b)

Figure 5.4 (*continued*)

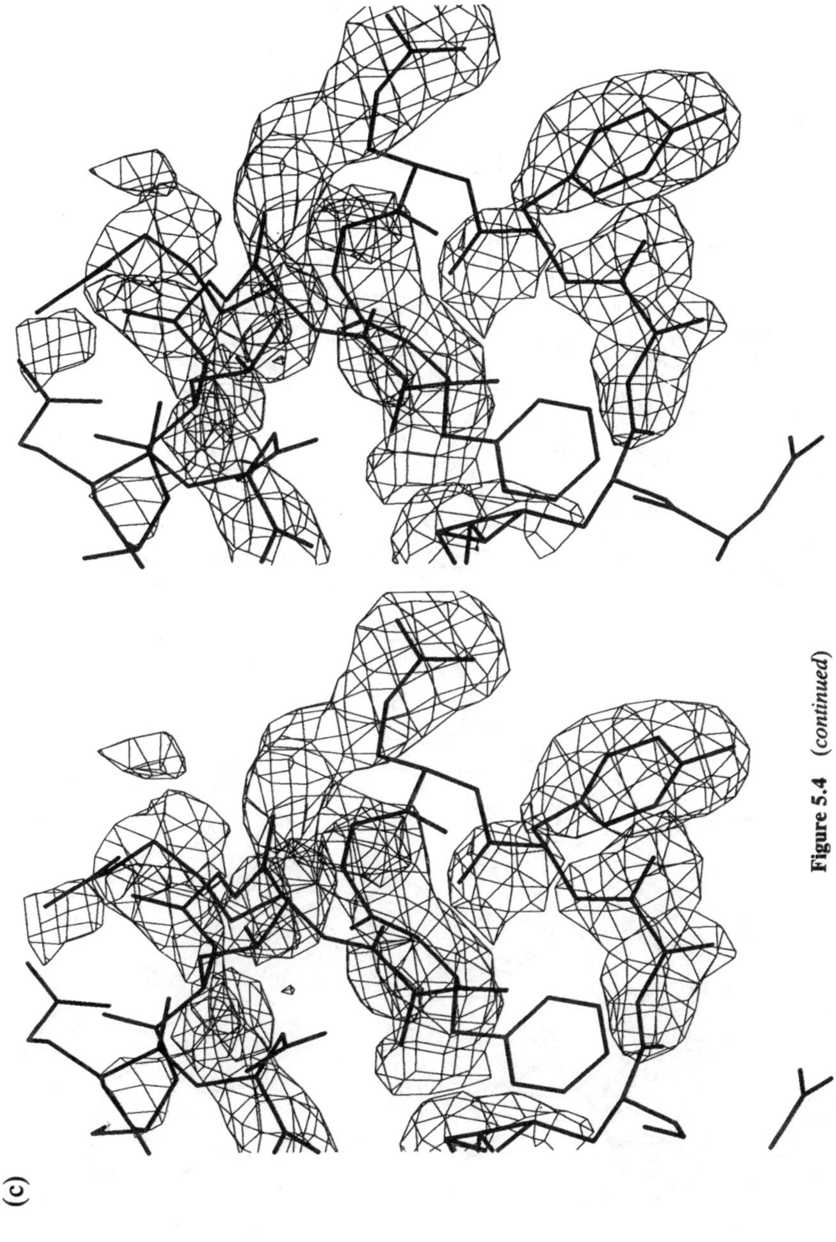

(c)

Figure 5.4 (*continued*)

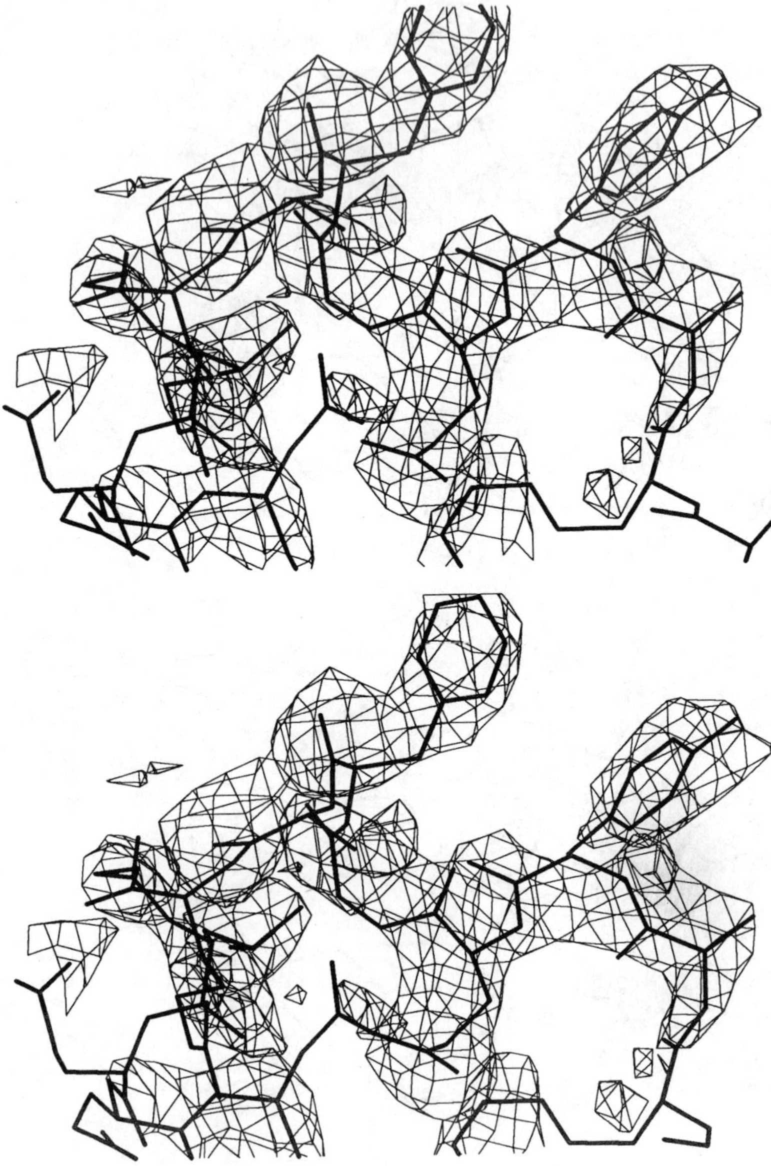

(d)

Figure 5.4 *(continued)*

of the final SA-refined structure (Figure 5.4(c)) clearly shows the density of the atomic positions of the correct structure (Figure 5.4(d)). In marked contrast, the electron density map of the structure obtained by conventional restrained least-squares refinement (Figure 5.4(b)) is difficult to interpret. Thus, the quality of SA-refined maps can simplify the model re-fitting in difficult regions, thus reducing the overall amount of human intervention during crystallographic refinement.

Indicators for the accuracy of refined structures are refined temperature factors (Kuriyan *et al.*, 1986) and the atomic RMS differences of several independently refined structures. In least-squares refinement, comparisons of independently refined structures underestimate the error if the refinement of one structure is based upon another refined model, since the second will remain near the local minimum reached during the first refinement round. Since SA refinement allows large movements determined by randomly assigned velocities, it provides a relatively unbiased estimate of the error in a structure. Figure 5.5 shows families of structures for two regions of aspartate aminotransferase that were obtained by cooling the structure a few times during a 3 ps heating stage. The α-helical region (Figure 5.5(a)) is well converged and shows RMS differences of 0.3 Å for backbone atoms between individual structures. This error in atomic coordinates is expected considering the high resolution limit of 2.8 Å for the aspartate aminotransferase diffraction data (Luzatti, 1952). In marked contrast, residues 223–229 (Figure 5.5(b)) show larger variations in that the RMS differences are 0.6 Å for backbone atoms. As was noted above, this region has been erroneously built and convergence is not achieved during SA refinement. From this result it was concluded that regions that exhibit large RMS differences between independently refined structures are either disordered or erroneously built.

Indicators for the correctness of the refined structure are deviations of local geometry from ideality, the fit of the model to various electron density maps, and the lack of large peaks in difference electron density maps. In spite of the success of the SA methodology, the importance of manual examination of the electron density maps after SA refinement has to be emphasized. This is essential for the placement of surface sidechains and solvent molecules, and for checking regions of the protein where large deviations from idealized geometry occur. Figure 5.6 illustrates a problem that occurred during SA refinement of influenza virus haemagglutinin (Weis *et al.*, 1990). A poorly defined tryptophan sidechain moved into strong density belonging to *N*-linked carbohydrate that was not included in the first round of SA refinement (Figure 5.6(a)). The model was manually rebuilt in this region, and the missing carbohydrate added. In subsequent rounds of annealing, proper model geometry and fit to the electron density map were maintained (Figure 5.6(b)). The problem in this region of haemagglutinin was produced because molecular dynamics can move

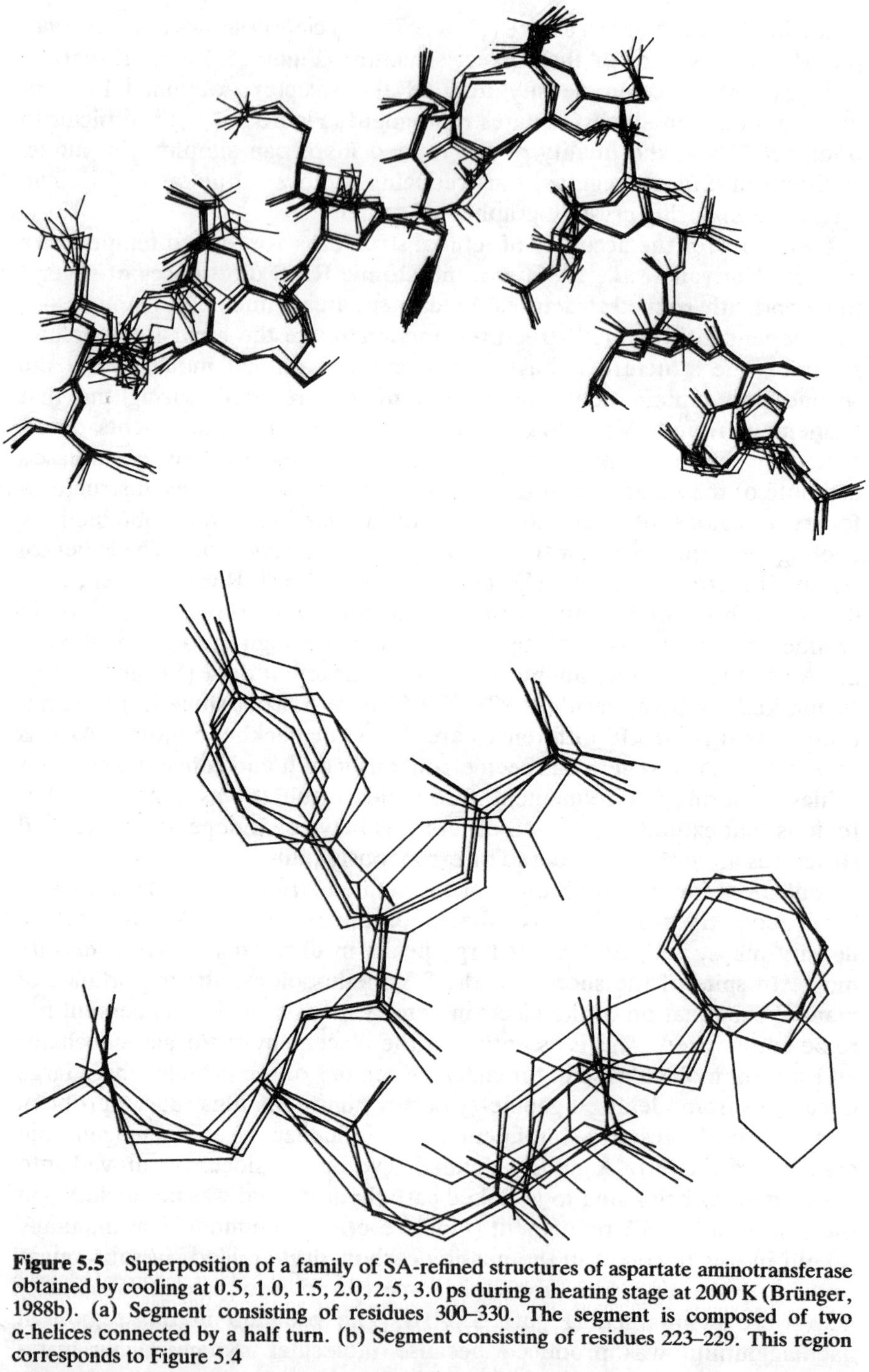

Figure 5.5 Superposition of a family of SA-refined structures of aspartate aminotransferase obtained by cooling at 0.5, 1.0, 1.5, 2.0, 2.5, 3.0 ps during a heating stage at 2000 K (Brünger, 1988b). (a) Segment consisting of residues 300–330. The segment is composed of two α-helices connected by a half turn. (b) Segment consisting of residues 223–229. This region corresponds to Figure 5.4

atoms far enough to compensate at least partially for missing model atoms, producing somewhat ambiguous difference electron density maps.

SA refinement is most useful when the initial model is relatively crude. Given an already well-refined model, SA refinement offers little advantage over conventional methods, with the possible exception of providing information about the accuracy of the refined structure. However, when only a crude model is available, SA refinement is able to reduce greatly the amount of human intervention. The initial model can be as crude as one that is obtained by automatic building based on C^{α} positions alone (A. Jones, personal communication).

COMBINED MOLECULAR REPLACEMENT AND PATTERSON REFINEMENT

In macromolecular crystallography, the initial determination of phases by molecular replacement (MR) (Hoppe, 1957; Rossmann and Blow, 1962; Huber, 1965; Rossmann, 1972; Lattman, 1985) is often attempted if the structure of a similar or homologous macromolecule is known ('search model'). MR involves the placement (i.e. rotation and translation) of the search model in the unit cell of the target crystal in order to obtain the best agreement between calculated and observed diffraction data. The optimally placed search model is used to obtain initial phases for crystallographic structure fitting and refinement.

Progress in obtaining approximate three-dimensional search models of macromolecules from other information suggests an increased use of MR to solve crystal structures. For instance, the data base of known protein sequences and protein structures is growing rapidly. Techniques for aligning sequences such as consensus templates have been developed in order to recognize very distantly related proteins or protein domains and to carry out model building on the basis of the known protein structures. Another example is the determination of three-dimensional structures of small proteins and nucleic acids based on interproton distances obtained from NMR NOESY experiments (Wüthrich, 1986). The search models obtained by these methods are often approximate, which makes MR very difficult, if not impossible. In fact, most successful MR cases reported involve search models with a backbone atomic RMS difference of less than 1 Å from the target structure.

If there is one molecule in the crystallographic asymmetric unit, then three positional and three angular parameters fully describe the placement of the search model in the unit cell of the target crystal. Rossmann and Blow (1962) realized that this six-dimensional search can be reduced to a sequence of a three-dimensional angular search using a 'rotation function' followed by a three-dimensional positional search using a 'translation

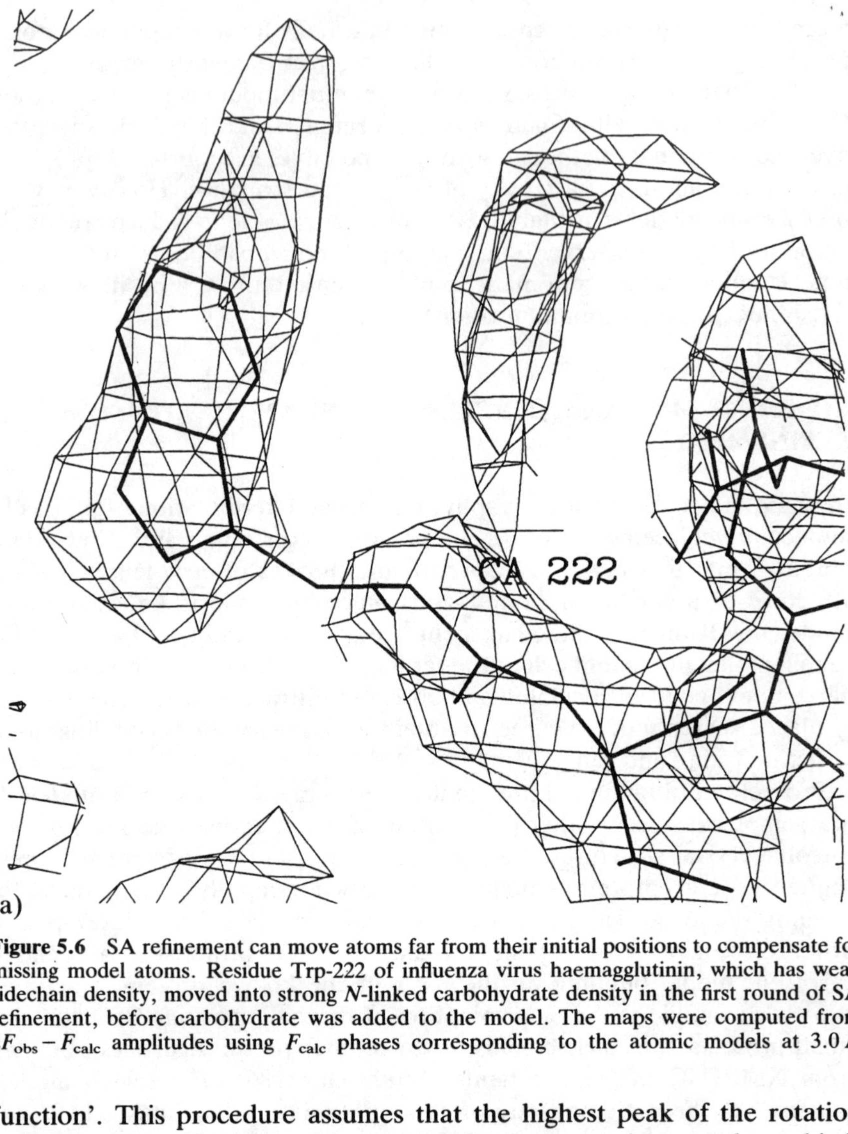

(a)

Figure 5.6 SA refinement can move atoms far from their initial positions to compensate for missing model atoms. Residue Trp-222 of influenza virus haemagglutinin, which has weak sidechain density, moved into strong *N*-linked carbohydrate density in the first round of SA refinement, before carbohydrate was added to the model. The maps were computed from $2F_{obs} - F_{calc}$ amplitudes using F_{calc} phases corresponding to the atomic models at 3.0 Å

function'. This procedure assumes that the highest peak of the rotation function yields the correct orientation. Examples are known where this is not true. As a result of advances in computer technology, multi-dimensional search strategies with more than three parameters are no longer beyond available computational resources.

Recently it was proposed to generalize conventional MR by allowing certain coordinates of the search model to vary in a neighbourhood N of the initial positions in addition to the overall orientation and position of the molecule (Brünger, 1990). In the limiting case of $N = 0$, the method is identical to conventional MR, and in the limiting case of $N \gg 0$ the method

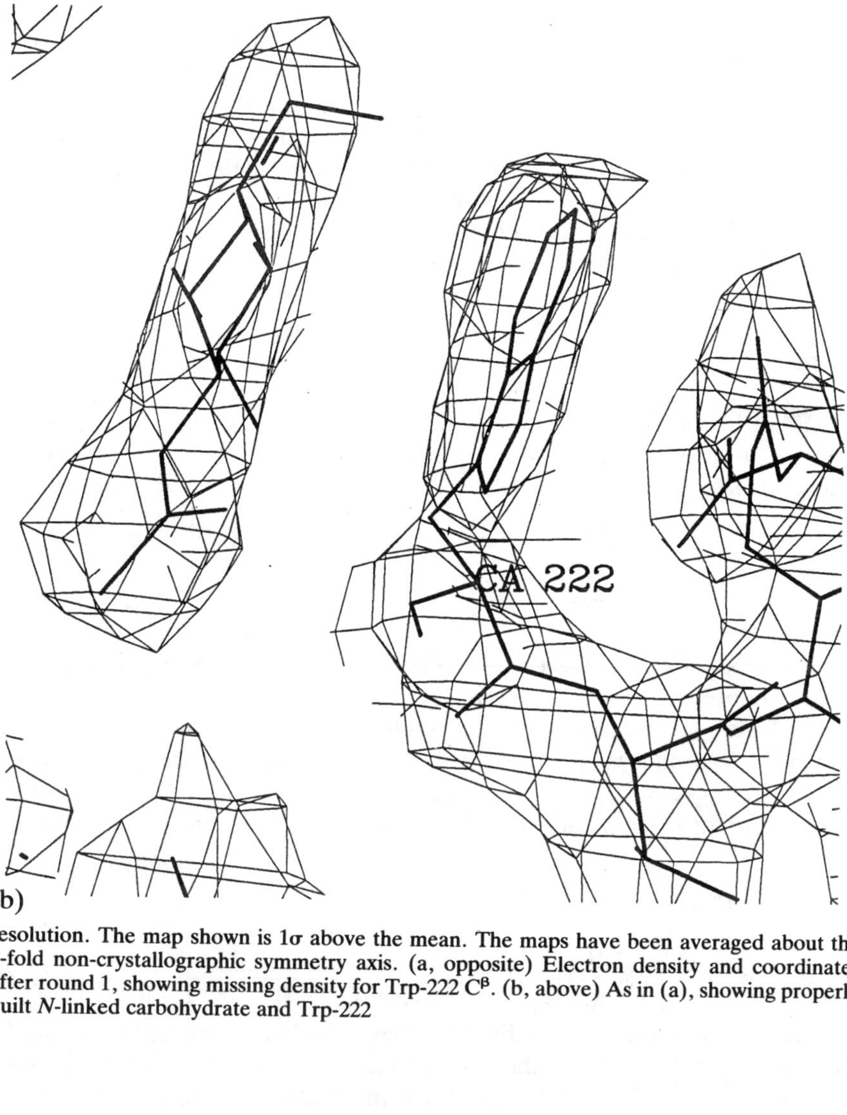

(b)

resolution. The map shown is 1σ above the mean. The maps have been averaged about the 3-fold non-crystallographic symmetry axis. (a, opposite) Electron density and coordinates after round 1, showing missing density for Trp-222 C^{β}. (b, above) As in (a), showing properly built *N*-linked carbohydrate and Trp-222

is equivalent to the crystallographic phase problem; so far we have focused on applications where N is small and falls within the radius of convergence of conjugate gradient minimization. It was shown that coordinates (either individual atomic coordinates or coordinates of several rigid bodies) can be refined against a target function that consists of an empirical energy function combined with a 'Patterson' energy term describing the discrepancy between the observed and calculated Patterson functions. This is the third example of this review that involves a hybrid energy function (Equation 5.1). The method can be used to refine the search model prior to translation searches, thus increasing the significance of both the correct orientation of the rotation function as well as that of the translation function. This method has been termed PC refinement. A combined MR and PC refinement strategy was proposed. First, a conventional RF is carried out. All sampled orientations are sorted with respect to their RF value and then a large number of the highest peaks are selected for PC refinements. Finally, the PC-refined search models with the highest correlation coefficients are used for conventional translation searches.

Patterson Energy Term

MR can be viewed as a non-linear optimization problem with the aim of maximizing the standard linear correlation coefficient PC (Fujinaga and Read, 1987), given by

$$\mathrm{PC}(\Omega,T) = \frac{\langle |E_{\mathrm{obs}}|^2 |E_{\mathrm{m}}(\Omega,T)|^2 - \langle |E_{\mathrm{obs}}|^2\rangle \langle |E_{\mathrm{m}}(\Omega,T|^2\rangle\rangle\rangle}{\sqrt{\langle |E_{\mathrm{obs}}|^4 - \langle |E_{\mathrm{obs}}|^2\rangle^2\rangle \langle |E_{\mathrm{m}}(\Omega,T)|^4 - \langle |E_{\mathrm{m}}(\Omega,T)|^2\rangle^2\rangle}} \tag{5.27}$$

The symbols $\langle\rangle$ denote an averaging over the set of observed reflections. E_{obs} denote the normalized observed structure factors and $E_{\mathrm{m}}(\Omega,T)$ denote the normalized structure factors of the search model oriented according to the rotation matrix Ω and positioned according to the translation vector T. The importance of the work of Brünger (1990) consists of introducing additional degrees of freedom into Equation (5.27), i.e.

$$\mathrm{PC}(\boldsymbol{r},\Omega,T) = \frac{\langle |E_{\mathrm{obs}}|^2 |E_{\mathrm{m}}(\boldsymbol{r},\Omega,T)|^2 - \langle |E_{\mathrm{obs}}|^2\rangle \langle |E_{\mathrm{m}}(\boldsymbol{r},\Omega,T)|^2\rangle\rangle}{\sqrt{\langle |E_{\mathrm{obs}}|^4 - \langle |E_{\mathrm{obs}}|^2\rangle^2\rangle \langle |E_{\mathrm{m}}(\boldsymbol{r},\Omega,T)|^4 - \langle |E_{\mathrm{m}}(\boldsymbol{r},\Omega,T)|^2\rangle^2\rangle}} \tag{5.28}$$

where $\boldsymbol{r}$ denote atomic coordinates or rigid body coordinates of the search model. If a single copy of the search model is oriented according to Ω and placed in a triclinic unit cell identical in geometry to that of the crystal, Equation (5.28) becomes independent of the translation T. It is then possible to refine the search model coordinates against PC prior to translation searches by defining the Patterson energy term $E_{\mathrm{exp}}^{\mathrm{PC}}$:

$$E_{\mathrm{exp}}^{\mathrm{PC}}(\boldsymbol{r},\Omega) = 1 - \mathrm{PC}(\boldsymbol{r},\Omega) \tag{5.29}$$

Applications

First results of PC refinement are encouraging. A number of computer studies were carried out by Brünger (1990) in order to evaluate the radius of convergence of PC refinement. A search model of crambin with a 2 Å backbone atomic RMS difference from the crystal structure that was obtained from an NMR structure determination using simulated data (Brünger *et al.*, 1986; Clore *et al.*, 1986a) failed to provide the correct orientation when using a rotation function or a six-dimensional search. PC refinements of the search model in several orientations resulted in the correct orientation having the lowest value of E_{total}. This enabled us to solve the crambin crystal structure starting with the PC-refined search model. In an application to myoglobin we showed that rigid-group PC refinement of the orientations of the eight α-helices has an approximate radius of convergence of 13°. The crystal structure of myoglobin was made inaccurate by tilting the eight α-helices artificially by 13° around independent axes. This search model failed to provide the correct orientation when using a conventional rotation function (Figure 5.7(a)) or a six-dimensional search. PC refinements uniquely determined the correct overall orientation of the myoglobin search model by returning the α-helices to their original orientations (Figure 5.7(b)).

Recently, we have been able to solve an Fab fragment of a monoclonal antibody with bound hapten (Leahy *et al.*, 1988) by molecular replacement combined with PC refinement. The search model consisted of the HyHEL-5 Fab of the Fab–lysozyme complex of Sheriff *et al.* (1987) with sequence similarities of 84%, 100%, 40%, and 100% for the variable domain of the light chain, constant domain of the light chain, variable domain of the heavy chain, and the constant domain of the heavy chain respectively. The correct orientation was not significantly above error peaks in the rotation function. PC refinements of the rigid body coordinates of the four domains using the most likely orientations uniquely determined the correct orientation. Furthermore, the significance of the translation function was improved when the PC-refined search model was used. The *R* factor of the correctly placed Fab search model was 42% at 2.8 Å resolution. SA refinement reduced the *R* factor to 23% in one round with relatively minor adjustments necessary to refine the structure fully.

CONCLUDING REMARKS

Energy minimization, molecular dynamics and SA have become powerful tools for the structural biologist working with X-ray crystallographic or NMR spectroscopic data. The primary application is for the refinement of macromolecular structures based on X-ray diffraction or NMR interproton distance data. However, the methods can also play a role for structure determination. For instance, molecular dynamics can yield correctly folded structures using interproton distance restraints starting with extended

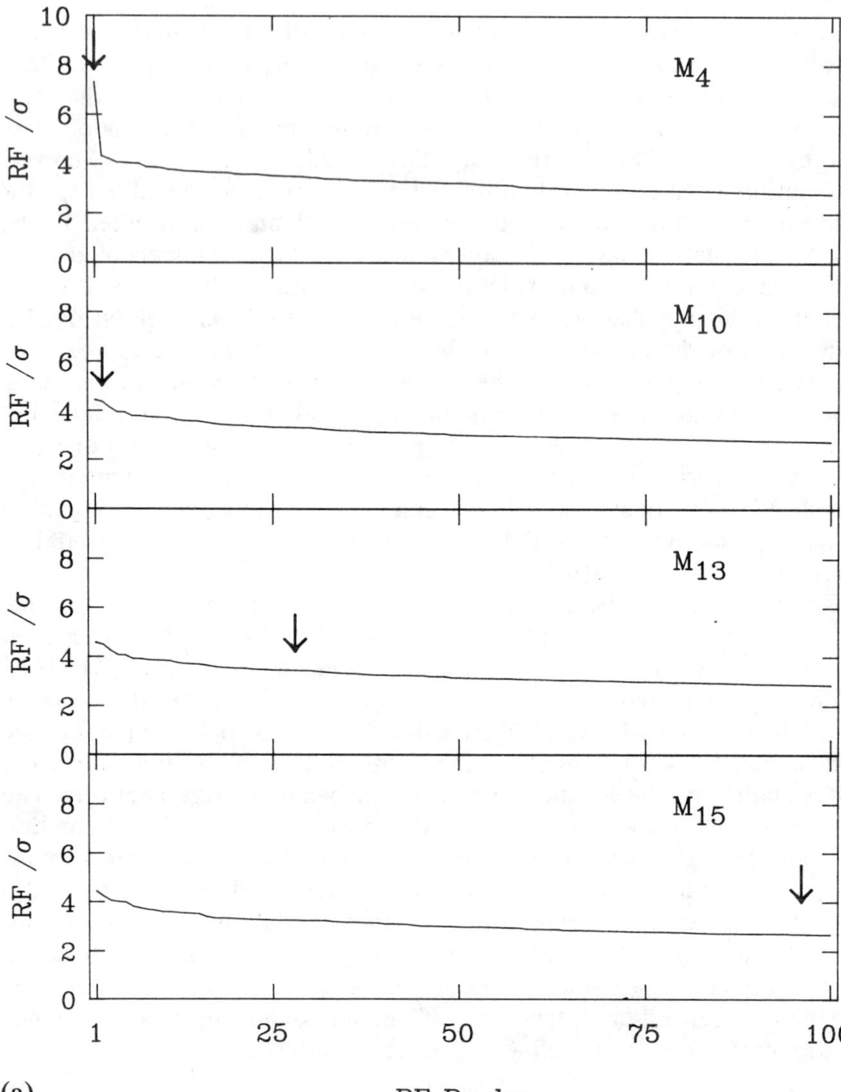

Figure 5.7 (a) Rotation searches for myoglobin search models M_4, M_{10}, M_{13}, and M_{15} at 15–3 Å resolution with the eight α-helices tilted by 4°, 10°, 13°, and 15°, respectively. Shown are the values of the 100 highest peaks of a conventional rotation function (RF) sorted by peak height. The correct orientation is indicated by an arrow. Details of the RF are described

chains or random arrays of atoms. Another example is molecular replacement to obtain initial phases for single-crystal diffraction data, where PC refinement can improve the signal-to-noise ratio of rotation and translation functions.

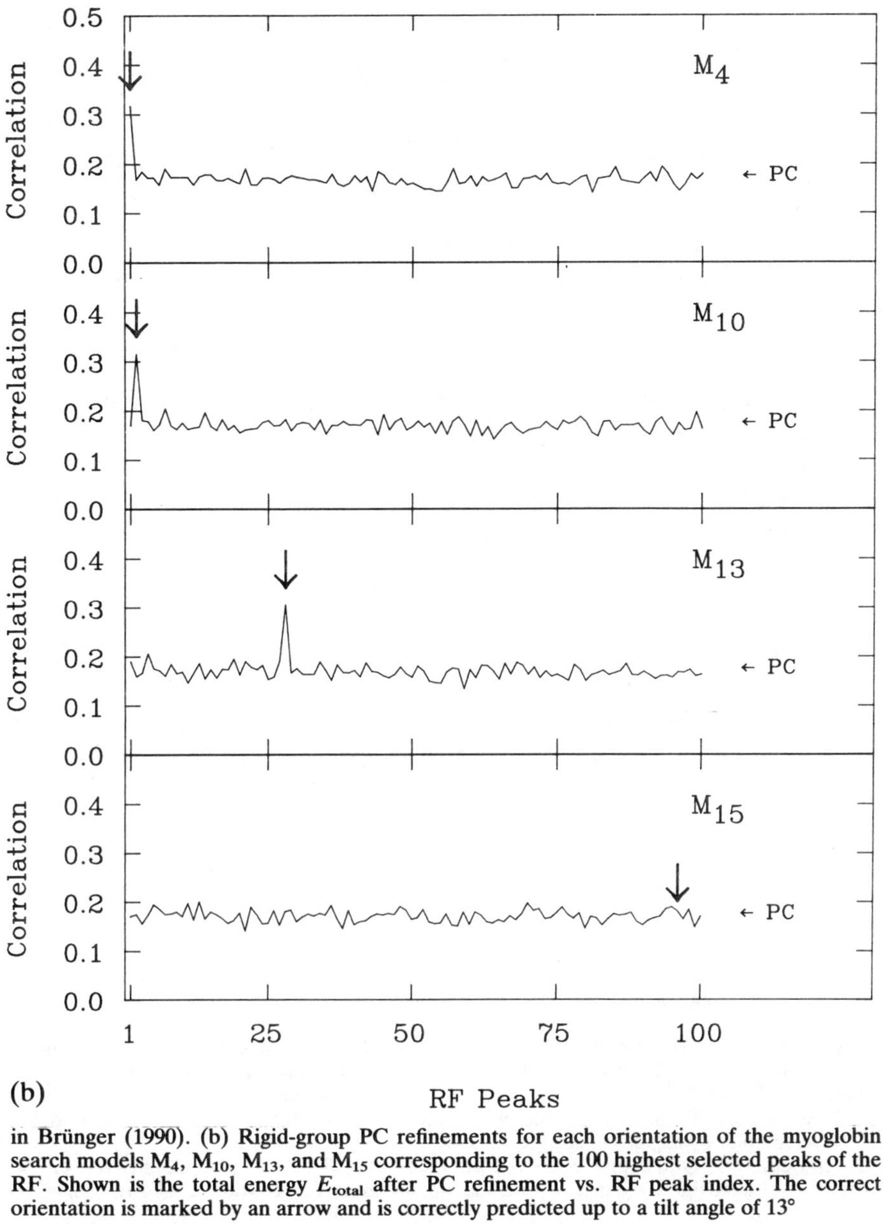

(b)

in Brünger (1990). (b) Rigid-group PC refinements for each orientation of the myoglobin search models M_4, M_{10}, M_{13}, and M_{15} corresponding to the 100 highest selected peaks of the RF. Shown is the total energy E_{total} after PC refinement vs. RF peak index. The correct orientation is marked by an arrow and is correctly predicted up to a tilt angle of 13°

The hybrid energy function approach (Equation 5.1) is general and it can be applied to any situation where fitting of an atomic model to measured data is required. It is conceivable that it will be applied to other types of diffraction data, such as fibre diffraction or electron diffraction, or to other types of distance measurements obtained from spectroscopic techniques.

Several issues of macromolecular refinement were omitted in this review. For example, in X-ray crystallography it is still not fully understood why, regardless of the refinement method used, the *R* factors of refined macromolecular structures (typically 20–10%) do not reach *R* factors that are obtained for small molecule structures. Presumably, this discrepancy is at least partially due to inadequate or incomplete treatment of temperature factors, anisotropy, alternative conformations, and structured and bulk solvent. The relationship between temperature factors, alternative conformations, and anisotropy has been studied through molecular dynamics simulations by Kuriyan *et al.* (1986). These studies have provided insight into the accuracy of crystal structures.

In the area of solution NMR spectroscopy, spin-diffusion and motional averaging will certainly have to be addressed to obtain more accurate structures. The complete relaxation matrix treatment (Borgias and James, 1988; Yip and Case, 1989) is the first step towards a solution NMR refinement technique where calculated NOE intensities are fitted to observed NOE intensities.

ACKNOWLEDGEMENTS

I wish to thank my colleagues who have directly or indirectly in discussions contributed to the work reviewed in this paper. They include G. M. Clore, J. W. Erickson, A. M. Gronenborn, R. Huber, M. Karplus, J. Kuriyan, M. Nilges, G. A. Petsko and W. I. Weis.

REFERENCES

Acharya, R., Fry, E., Stuart, D., Fox, G., Rowlands, D. and Brown, F. (1989). The three-dimensional structure of foot-and-mouth disease virus at 2.9 Å resolution. *Nature*, **337**, 709–716

Arnold, E. and Rossmann, M. G. (1988). The use of molecular-replacement phases for the refinement of the human rhinovirus 14 structure. *Acta Crystallogr. A*, **44**, 270–282

Bassolino, D., Hirata, F., Kitchen, D. B., Kominos, D., Pardi, A. and Levy, R. M. (1988). Determination of protein structures in solution using NMR data and impact. *Int. J. Supercomput. Appl.*, **2**, 41–61

Berendsen, H. J. C., Postma, J. P. M., van Gunsteren, W. F., DiNola, A. and Haak, J. R. (1984). Molecular dynamics with coupling to an external bath. *J. Chem. Phys.*, **81**, 3684–3690

Billeter, M., Havel, T. F. and Wüthrich, K. (1987). The ellipsoid algorithm as a method for the determination of polypeptide conformations from experimental distance constraints and energy minimization. *J. Comput. Chem.*, **8**, 132–141

Borgias, G. A. and James, T. L. (1988). COMATOSE, a method for constrained refinement of macromolecular structure based on two-dimensional nuclear Overhauser effect spectra. *J. Magn. Reson.*, **79**, 493–512

Bounds, D. G. (1987). New optimization methods from physics and biology. *Nature*, **329**, 215–219

Braun, W. (1987). Distance geometry and related methods for protein structure determination from NMR data. *Q. Rev. Biophys.*, **19**(3/4), 115–157

Braun, W. and Go, N. (1985). Calculation of protein conformations by proton–proton distance constraints. A new efficient algorithm. *J. Mol. Biol.*, **186**, 611–626

Brooks, B. R., Bruccoleri, R. E., Olafson, B. D., States, D. J., Swaminathan, S. and Karplus, M. (1983). CHARMM: a program for macromolecular energy, minimization, and dynamics calculations. *J. Comput. Chem.*, **4**, 187–217

Brooks, C. L. III, Karplus, M. and Pettitt, B. M. (1988). In *Proteins: a Theoretical Perspective of Dynamics, Structure and Thermodynamics*. Wiley, New York

Brünger, A. T. (1988a). *X-PLOR Manual*, Version 1.5. Yale University

Brünger, A. T. (1988b). Crystallographic refinement by simulated annealing: application to a 2.8 Å resolution structure of aspartate aminotransferase. *J. Mol. Biol.*, **203**, 803–816

Brünger, A. T. (1989). A memory-efficient fast fourier transformation algorithm for crystallographic refinement on supercomputers. *Acta Crystallogr. A*, **45**, 42–50

Brünger, A. T. (1990). Extension of molecular replacement: a new search strategy based on Patterson correlation refinement. *Acta Crystallogr. A*, **46**, 46–57

Brünger, A. T., Brooks, C. L. and Karplus, M. (1984). Stochastic boundary conditions for molecular dynamics simulations of ST2 wastes. *Chem. Phys. Lett.*, **105**, 495–500

Brünger, A. T., Clore, G. M., Gronenborn, A. M. and Karplus, M. (1986). Three-dimensional structures of proteins determined by molecular dynamics with interproton distance restraints: application to crambin. *Proc. Natl. Acad. Sci. USA*, **83**, 3801–3805

Brünger, A. T., Clore, G. M., Gronenborn, A. M. and Karplus, M. (1987a). Solution conformations of a human growth hormone releasing factor: comparison of the restrained molecular dynamics and distance geometry methods for a system without long-range distance data. *Protein Eng.*, **1**, 399–406

Brünger, A. T., Karplus, M. and Petsko, G. A. (1989a). Crystallographic refinement by simulated annealing: application to crambin. *Acta Crystallogr. A*, **45**, 50–61

Brünger, A. T., Krukowski, A. and Erickson, J. W. (1990). *Acta Crystallogr. A*, **46**, 585–593

Brünger, A. T., Kuriyan, J. and Karplus, M. (1987b). Crystallographic *R* factor refinement by molecular dynamics. *Science*, **235**, 458–460

Champness, J. N., Bloomer, A. C., Bricogne, G., Butler, P. J. G. and Klug, A. (1976). The structure of the protein disk of tobacco mosaic virus to 5 Å resolution. *Nature*, **259**, 20–24

Clore, G. M. and Gronenborn, A. M. (1989). Determination of three-dimensional structures of proteins and nucleic acids in solution by nuclear magnetic resonance spectroscopy. *CRC Crit. Rev. Biochem.*, **24**, 479–564

Clore, G. M., Gronenborn, A. M., Brünger, A. T. and Karplus, M. (1985). Solution conformation of a heptadecapeptide comprising the DNA binding helix F of the cyclic AMP receptor protein of *E. coli.* Combined use of ^{1}H NMR and restrained molecular dynamics. *J. Mol. Biol.*, **186**, 435–455

Clore, G. M., Brünger, A. T., Karplus, M. and Gronenborn, A. M. (1986a). Application of molecular dynamics with interproton distance restraints to three-dimensional protein structure determination: a model study of crambin. *J. Mol. Biol.*, **191**, 523–551

Clore, G. M., Nilges, M., Sukumaran, D. K., Brünger, A. T., Karplus, M. and Gronenborn, A. M. (1986b). The three-dimensional structure of alpha 1-purothionin in solution: combined use of nuclear magnetic resonance, distance geometry and restrained molecular dynamics. *EMBO J*, **5**, 2729–2735

Clore, G. M., Oschkinat, H., McLaughlin, L. W., Benseler, F., Happ, C. S., Happ, E. and Gronenborn, A. M. (1988). Refinement of the solution structure of the DNA dodecamer 5′d(CGCGPATTCGCG)$_2$ containing a stable purine–thymine base pair: combined use of nuclear magnetic resonance and restrained molecular dynamics. *Biochemistry*, **27**, 4185–4197

Crippen, G. M. (1977). Conformational analysis by energy embedding. *J. Comput. Phys.*, **24**, 96–107

Driscoll, P. C., Gronenborn, A. M., Beress, L. and Clore, G. M. (1989). Determination of the three-dimensional solution structure of the antihypertensive and antiviral protein

BDS-I from the sea anemone *Anemonia sulcata*: a study using nuclear magnetic resonance and hybrid distance geometry–dynamical simulated annealing. *Biochemistry*, **28**, 2188–2198

Ernst, R. R., Bodenhausen, G. and Wokaun, A. (1986). In *Principles of Nuclear Magnetic Resonance in One and Two Dimensions*. Clarendon, Oxford

Fujinaga, M. and Read, R. J. (1987). Experiences with a new translation-function program. *J. Appl. Crystallogr.*, **20**, 517–521

Fujinaga, M., Gross, P. and van Gunsteren, W. F. (1989). Testing the method of crystallographic refinement using molecular dynamics. *J. Appl. Crystallogr.*, **22**, 1–8

Gouaux, J. E. and Lipscomb, W. N. (1988). Three-dimensional structure of carbonyl phosphate and succinate bound to aspartate carbamoyl transferase. *Proc. Natl. Acad. Sci. USA*, **85**, 4205–4208

Gronenborn, A. M. and Clore, G. M. (1989). Analysis of the relative contributions of the nuclear Overhauser interproton distance restraints and the empirical energy function in the calculation of oligonucleotide structures using restrained molecular dynamics. *Biochemistry*, **28**, 5978–5984

Havel, T. F. (1986). *DISGEO*, Quantum Chemistry Program Exchange Program No. 507. Indiana University

Havel, T. F. and Wüthrich, K. (1985). An evaluation of the combined use of nuclear magnetic resonance and distance geometry for determination of protein conformations in solution. *J. Mol. Biol.*, **182**, 281–294

Hendrickson, W. A. (1985). Stereochemically restrained refinement of macromolecular structures. *Methods Enzymol.*, **115**, 252–270

Hendrickson, W. A. and Lattman, E. E. (1970). Representation of phase probability distributions for simplified combination of independent phase information. *Acta Crystallogr. B*, **26**, 136–143

Hermans, J., Berendsen, H. J. C., van Gunsteren, W. F. and Postma, J. P. M. (1984). A consistent empirical potential for water–protein interactions. *Biopolymers*, **23**, 1513–1518

Hoppe, W. (1957). Die ‹Faltmolekulmethode›—eine neue Methode für Bestimmung der Kristallstruktur bei ganz oder Teilweise Bekannter. Molekulstruktur. *Acta Cryst.*, **10**, 750–751

Huber, R. (1965). Die automatisierte Faltmolekülmethode. *Acta Cryst. A*, **19**, 353–356

Hyberts, G., Märki, W. and Wagner, G. (1987). Stereospecific assignments of side-chain protons and characterization of torsion angles in Eglin c. *Eur. J. Biochem.*, **164**, 625–635

International Tables for X-ray Crystallography (1974). Ibers, J. and Hamilton, W. C. (Eds.), Vol. 4. Kynoch, Birmingham

Jack, A. and Levitt, M. (1978). Refinement of large structures by simultaneous minimization of energy and *R* factor. *Acta Crystallogr. A*, **34**, 931–935

Jones, E. Y., Stuart, D. I. and Walker, N. P. C. (1989). Structure of tumor necrosis factor. *Nature*, **338**, 225–228

Jones, T. A. (1978). A graphics model building and refinement system for macromolecules. *J. Appl. Crystallogr.*, **11**, 268–272

Kaptein, R., Zuiderweg, E. R. P., Scheek, R. M., Boelens, R. and van Gunsteren, W. F. (1985). A protein structure from NMR data. lac repressor headpiece. *J. Mol. Biol.*, **182**, 179–182

Karplus, M. (1959). Contact electron-spin coupling of nuclear magnetic moments. *J. Chem. Phys.*, **30**, 11–15

Kirkpatrick, S., Gelatt, C. D. Jr. and Vecchi, M. P. (1983). Optimization by simulated annealing. *Science*, **220**, 671–680

Kline, A. D., Braun, W. and Wüthrich, K. (1988). Determination of the complete three-dimensional structure of the α-amylose inhibitor Tendamistat in aqueous solution by nuclear magnetic resonance and distance geometry. *J. Mol. Biol.*, **204**, 675–724

Konnert, J. H. and Hendrickson, W. A. (1980). A restrained-parameter thermal-factor refinement procedure. *Acta Crystallogr. A*, **36**, 344–349

Kuriyan, J. Petsko, G. A., Levy, R. M. and Karplus, M. (1986). Effect of anisotropy and anharmonicity on protein crystallographic refinement. *J. Mol. Biol.*, **190**, 227–254

Kuriyan, J., Brünger, A. T., Karplus, M. and Hendrickson, W. A. (1989). X-ray refinement of protein structures by simulated annealing: test of the method on myohemerythrin. *Acta Crystallogr.* A, **45**, 396–409

Lattman, E. E. (1985). Use of the rotation and translation functions. *Methods Enzymol.*, **115**, 55–77

Leahy, D. J., Hynes, T. R., McConnell, H. M. and Fox, R. O. (1988). Crystallization of an anti-2,2,6,6-tetramethyl-1-piperidinyloxy-dinitrophenyl monoclonal antibody Fab fragment with and without bound hapten. *J. Mol. Biol.*, **203**, 829–830

Lee, M. S., Gippert, G. P., Soman, K. V., Case, D. A. and Wright, P. E. (1989). Three-dimensional solution structure of a single zinc finger DNA-binding domain. *Science*, **245**, 635–637

Lifson, S. and Stern, P. (1982). Born–Oppenheimer energy surfaces of similar molecules: interrelations between bond lengths, bond angles, and frequencies of normal vibrations in alkanes. *J. Chem. Phys.*, **77**, 4542–4550

Luzatti, V. (1952). Traitement statistique des erreurs dans la determination des structures cristallines. *Acta Crystallogr.*, **5**, 802–810

McCammon, J. A. and Harvey, S. C. (1987). *Dynamics of Proteins and Nucleic Acids*. Cambridge University Press, Cambridge

Metropolis, N., Rosenbluth, M., Rosenbluth, A., Teller, A. and Teller, E. (1953). Equation of state calculations by fast computing machines. *J. Chem. Phys.*, **21**, 1087–1092

Navia, M. A., Fitzgerald, P. M. D., McKeever, B. M., Leu, C.-T., Heimbach, J. C., Herber, W. K., Sigal, I. S., Darke, P. L. and Springer, J. P. (1989). Three-dimensional structure of aspartyl protease from human immunodeficiency virus HIV-1. *Nature*, **337**, 615–620

Némethy, G., Pottie, M. S. and Scheraga, H. A. (1983). Energy parameters in polypeptides. 9. Updating of geometrical parameters, nonbonded interactions, and hydrogen bond interactions for the naturally occurring amino acids. *J. Phys. Chem.*, **87**, 1883–1887

Nilges, M., Clore, G. M. and Gronenborn, A. M. (1988a). Determination of three-dimensional structures of proteins from interproton distance data by dynamical simulated annealing from a random array of atoms. *FEBS Lett.*, **239**, 129–136

Nilges, M., Clore, G. M. and Gronenborn, A. M. (1988b). Determination of three-dimensional structures of proteins from interproton distance data by hybrid distance geometry–dynamical simulated annealing calculations. *FEBS Lett.*, **239**, 317–324

Nilges, M., Gronenborn, A. M. and Clore, G. M. (1989). Structure determination from NMR conformational data by molecular dynamics calculations. *Proceedings of the 1989 Daresbury Study Weekend*, Science and Engineering Research Council, Daresbury Laboratory, Warrington

Nilsson, L. and Karplus, M. (1986). Empirical energy functions for energy minimization and dynamics of nucleic acids. *J. Comput. Chem.*, **7**, 591–616

Nilsson, L., Clore, G. M., Gronenborn, A. M., Brünger, A. T. and Karplus, M. (1986). Structure refinement of oligonucleotides by molecular dynamics with nuclear Overhauser effect interproton distance restraints: application to 5′d(C-G-T-A-C-G)$_2$. *J. Mol. Biol.*, **188**, 455–475

Northrup, S. H. and McCammon, J. A. (1980). Simulation methods for protein structure fluctuations. *Biopolymers*, **19**, 1001–1016

Oschkinat, H., Griesinger, C., Kraulis, P. J., Sorensen, O. W., Ernst, R. R., Gronenborn, A. M. and Clore, G. M. (1988). Three-dimensional NMR spectroscopy of a protein in solution. *Nature*, **332**, 374–376

Parker, M. W., Pattus, F., Tucker, A. D. and Tsernoglu, D. (1989). Structure of the membrane-pore-forming fragment of colicin A. *Nature*, **337**, 93–96

Pastor, R. W., Brooks, B. R. and Szabo, A. (1988). An analysis of the accuracy of Langevin and molecular dynamics algorithms. *Mol. Phys.*, **65**, 1409–1419

Press, W. H., Flannery, B. P., Teukolosky, S. A. and Vetterling, W. T. (1986). In *Numerical Recipes*. Cambridge University Press, Cambridge, pp. 498–546

Purisima, E. O. and Scheraga, H. A. (1986). An approach to the multiple-minima problem by relaxing dimensionality. *Proc. Natl. Acad. Sci. USA*, **83**, 2782–2786

Rees, M. and Lewis, G. (1983). Incorporation of experimental phases in a restrained least-squares refinement. *Acta Crystallogr. A*, **39**, 94–97

Rossmann, M. G. (Ed.) (1972). *The Molecular Replacement Method*, International Science Review No. 13. Gordon and Breach, New York

Rossmann, M. G. and Blow, D. M. (1962). The detection of sub-units within the crystallographic asymmetric unit. *Acta Crystallogr. A*, **15**, 24–31

Rossmann, M. G., Arnold, E., Erickson, J. W., Frankenberger, E. A., Griffith, J. P., Hecht, H. J., Johnson, J. E., Kamer, G., Luo, M., Mosser, A. G., Rueckert, R. R., Sherry, B. and Vriend, G. (1985). Structure of human common cold virus and functional relationship to other picornaviruses. *Nature*, **317**, 145–153

Sheriff, S., Silverton, E. W., Padlan, E. A., Cohen, G. H., Smith-Gill, S. J., Finzel, B. C. and Davies, D. R. (1987). Three-dimensional structure of an antibody–antigen complex. *Proc. Natl. Acad. Sci. USA*, **84**, 8075–8079

Sielecki, A. R., Hayakawa, K., Fujinaga, M., Murphy, M. E. P., Fraser, M., Muir, A. K., Carilli, C. T., Lewicki, J. A., Baxter, J. D. and James, M. N. G. (1989). Structure of recombinant human renin, a target for cardiovascular drugs, at 2.5 Å resolution. *Science*, **243**, 1346–1350

Smith, D. L., Ringe, D., Finlayson, W. L. and Kirsch, J. F. (1986). Preliminary X-ray data from aspartate aminotransferase from *Escherichia coli*. *J. Mol. Biol.*, **191**, 301–302

Stout, G. H. and Jensen, L. H. (1989). In *X-ray Structure Determination, a Practical Guide*. Wiley, New York, pp. 341–419

Stuart, D. and Artymiuk, P. (1985). The use of phase combination in crystallographic refinement: the choice of amplitude coefficients in combined syntheses. *Acta Crystallogr. A*, **40**, 713–716

Sussman, J. L., Holbrook, S. R., Church, G. M. and Kim, S. H. (1977). A structure-factor least-squares refinement procedure for macromolecular structure using constrained and restrained parameters. *Acta Crystallogr. A*, **33**, 800–804

Ten Eyck, L. F. (1973). Crystallographic fast Fourier transforms. *Acta Crystallogr. A*, **29**, 183–191

Tronrud, D. E., Ten Eyck, L. F. and Matthews, B. W. (1987). An efficient general-purpose least-squares refinement program for macromolecular structures. *Acta Crystallogr. A*, **43**, 489–501

Verlet, L. (1967). Computer 'experiments' on classical fluids. I. Thermodynamical properties of Lennard-Jones molecules. *Phys. Rev.*, **159**, 98–105

Watenpaugh, K. D. (1985). Overview of phasing by isomorphous replacement. *Methods Enzymol.*, **115**, 3–15

Weber, P. C., Ohlendorf, D. H., Wendoloski, J. J. and Salemme, F. R. (1989). Structural origins of high-affinity biotin binding to streptavidin. *Science*, **243**, 85–88

Weiner, S. J., Kollman, P. A., Nguyen, D. T. and Case, D. A. (1986). An all atom force field for simulations of proteins and nucleic acids. *J. Comput. Chem.*, **7**, 230–252

Weis, W., Brünger, A. T., Skehel, J. J. and Wiley, D. C. (1990). Refinement of the influenza virus haemagglutinin by simulated annealing. *J. Mol. Biol.*, **212**, 737–761

Wlodawer, A., Miller, M., Jaskólski, M., Sathyanarayana, B. K., Baldwin, E., Weber, I. T., Selk, J. M., Clawson, L., Schneider, J. and Kent, S. B. H. (1989). Conserved folding in retroviral proteases: crystal structure of a synthetic HIV-1 protease. *Science*, **245**, 616–621

Wüthrich, K. (1986). In *NMR of Proteins and Nucleic Acids*. Wiley, New York

Wüthrich, K. (1989). The development of nuclear magnetic resonance spectroscopy as a technique for protein structure determination. *Acc. Chem. Res.*, **22**, 36–44

Yip, P. and Case, D. A. (1989). A new method for refinement of macromolecular structures based on nuclear Overhauser effect spectra. *J. Magn. Reson.*, **83**, 643–648

Yoon, C., Prive, G. G., Goodsell, D. S. and Dickerson, R. E. (1988). Structure of an alternating-B DNA helix and its relationship to A-tract DNA. *Proc. Natl. Acad. Sci. USA*, **85**, 6332–6336

6
Applications of Free Energy Calculations

Mark A. Williams, Mansoor A. S. Saqi and Julia M. Goodfellow

INTRODUCTION

One of the most exciting developments in the field of computer simulation has come from techniques which allow estimates of free energies (Beveridge and Dicapua, 1989a, b; van Gunsteren, 1989). Most previous simulations of macromolecules have only involved calculation of the average potential energy of the system. As it is the free energy which controls the equilibrium behaviour of a system of particles, this is an important development which has opened up a range of new applications which are relevant to designing rational strategies for the development of new drugs and for protein engineering experiments.

The ability to estimate free energy changes is essential for understanding transformations of molecular systems in solution (Appendix 1). Under normal experimental conditions, of constant pressure and temperature, the Gibbs free energy (G) controls the state of a reaction and can be considered as a sum of an enthalpic and an entropic term. The change in free energy, ΔG, is given by

$$\Delta G = \Delta H - T\,\Delta S$$

where ΔH is the change in enthalpy, ΔS is the change in entropy and T is the temperature. A more ordered system has a lower entropy. Any estimation of free energy includes the important entropic effects which are known to be a major component in for example the hydrophobic interaction. The more usual potential energy is related to the enthalpic component of the free energy and so does not include these entropic effects.

To gain some understanding of the various effects contributing to any change in free energy, we can consider a reaction such as the binding of an inhibitor to an enzyme. Initially the inhibitor and enzyme molecules are well separated and surrounded by aqueous solvent. As the inhibitor and enzyme come together, many of these water molecules will be released into the bulk liquid. Thus there will be a gain of enzyme–inhibitor electrostatic and van der Waals interactions and a loss of interactions with water to take into account when calculating the change in potential energy.

In the bound state the enzyme and inhibitor are more ordered in the sense that their positions relative to each other are now fixed (whereas previously they could have been at any random relative position and oriented at any random angle to each other). This increase in order leads to a decrease in entropy (ΔS) and an increase in free energy, and is thus unfavourable. However, the formation of the complex is also accompanied by the release of some water molecules from their relatively ordered bound positions on the surface of the inhibitor and enzyme molecules into the much less ordered bulk liquid. This component of the reaction results in an increase in entropy (ΔS) and thus a decrease in free energy and hence this term is favourable to the reaction. Whether a reaction occurs or not depends on an intricate balance between all these enthalpic and entropic energy terms.

BACKGROUND

The calculation of free energies from computer simulations began in the 1970s but the application to macromolecular systems, such as enzyme–substrate binding, is a more recent development (Wong and McCammon, 1986). A shift of emphasis from calculating the absolute free energies of molecular systems to calculating the free energy differences between closely similar systems, coupled with advances in computers, has made studies of biologically important systems possible.

Several methods (such as thermodynamic integration, Widom's method, umbrella sampling, overlap ratio) were developed to study fairly simple model systems such as hard spheres, Lennard-Jones fluids, liquid nitrogen (Quirke, 1980) or liquid water (Mezei, 1982). More recently these methods of calculating free energy differences were extended and applied to small systems of chemical and biological interest. Such applications have included the mutation of Cl^- to Br^- (Lybrand *et al.*, 1985), and the transformation of methanol to ethane in aqueous solvent (Jorgensen and Ravimohan, 1985). The latter transformation involves a significantly larger change in the system than those previously studied and consequently it has become one of the standard 'test systems' for free energy algorithms.

The motivation for studying these small molecules is to develop the techniques so that they may be applied confidently to the study of macromolecular systems. A recent and comprehensive review of chemical and biochemical applications of free energy simulations (Beveridge and Dicapua, 1989b) concludes that we now have sufficient knowledge and experience of simple systems to begin to tackle complex biological problems. Such problems include enzyme–inhibitor binding, catalytic free energies, protein folding, DNA structure, and DNA–drug interactions.

THEORY AND COMPUTATIONAL PROCEDURES

Statistical mechanics relates macroscopic thermodynamic properties of a system to its individual microscopic states (Reif, 1965; Huang, 1969). For example, the mean energy of a system (U) is given by

$$U = \int \ldots \int E(q^{N})\, P(q^{N})\, dq^{N}$$

where $E(q^{N})$ is the energy of the system in one specific configuration q^{N}. $P(q^{N})$ is the Boltzmann factor which gives the probability of the configuration with energy $E(q^{N})$ where

$$P(q^{N}) = \exp[-E(q^{N})/kT]/Z$$

where Z is known as the partition function of the system as given by

$$Z = \int \ldots \int \exp[-E(q^{N})/kT]\, dq^{N}$$

The free energy, A, can be related to the partition function (see for example Mezei and Beveridge, 1986) by

$$\begin{aligned} A &= -kT \ln(Z/V^{N}) \\ &= kT \ln\langle \exp[E(q^{N})/kT] \rangle \end{aligned}$$

where $\langle\rangle$ implies an average over all configurations.

Thus, on the computer, we need to generate enough representative configurations in order that we can calculate realistic average values. Obtaining good sampling and estimating errors in the sampling for any system are major problems which have to be addressed for all types of simulations. The two main simulation methods that can be used to estimate free energy differences are molecular dynamics and Monte Carlo methods (as described in Chapter 1). Molecular dynamics explores the configurations by moving the atoms according to the forces acting between them (McCammon and Harvey, 1987). The Monte Carlo method moves atoms randomly but only accepts a new configuration with a probability related to the Boltzmann factor (Metropolis *et al.*, 1953), thus ensuring that configurations are generated with energies which give the correct equilibrium energy distribution.

One of the main problems with the computation of absolute free energies is that these simulation methods produce mostly low energy configurations, whereas the largest and consequently most important terms contributing to the free energy are those with high potential energy, i.e.

the configurations for which $\exp(E/kT)$ is large contribute most towards the average $\langle\exp(E/kT)\rangle$.

In order to overcome this sampling problem, Torrie and Valleau (1977) extended the range of sampling by using a weighted Boltzmann function as the acceptance criterion. They showed that the appropriate Boltzmann averages could still be recovered from a simulation with sampling carried out on the modified function. This 'umbrella sampling' technique has found many applications (Quirke, 1980; Mezei, 1982; Sussman *et al.*, 1985). However, an appropriate weighting function has to be chosen for each application and the choice of this is generally very difficult.

For many applications, it is not necessary to calculate an absolute free energy. Often, what is required is the free energy difference between two related states. Recently the free energy perturbation method has become the popular method for calculation of these free energy differences and it is implemented in the major computer packages for biomolecular simulations such as AMBER (Weiner and Kollman, 1981) CHARMM (Brooks *et al.*, 1983), GROMOS (Hermans *et al.*, 1984) and BOSS (W. L. Jorgensen, Purdue University, Indiana).

The free energy perturbation method can be illustrated by considering the difference in free energy of solvation of two solute molecules. We choose one solute molecule (and surrounding solvent) to be the reference system and generate equilibrium configurations of this reference system calculating the potential energy of each configuration (E_0). For each of these configurations we replace the reference solute by the other 'mutant' solute and recalculate the energy E_1. The free energy difference is then given by

$$\Delta A = -kT \ln\langle\exp[(E_0 - E_1)/kT]\rangle_0$$

where $\langle\rangle_0$ denotes that the average is taken over configurations of the reference state. Although this expression is exact, it is only possible to obtain accurate averages if the difference between two solutes is small.

If the reference and mutant states differ significantly, the free energy difference is calculated in a series of steps. Several hybrid intermediate non-physical states are created with the aid of a coupling parameter. This coupling parameter is used to mutate bond lengths, non-bonded interaction parameters and partial charges on the solute atoms. The reference system is thus transformed to the mutant through a number of steps and the total free energy change, ΔA_λ is the sum of all the free energy differences ΔA_λ for these 'thermodynamic windows' between states $\lambda = 0$ and $\lambda = 1$:

$$\Delta A = \sum_{\lambda=0}^{\lambda=1} \Delta A_\lambda$$

It is also usual to implement double-wide sampling in these simulations as at each step λ_i we can calculate both the forward $\Delta A_{\lambda_{i \ldots i+1}}$ and the backward $\Delta A_{\lambda_{i \ldots i-1}}$ free energies simultaneously. The differences between the forward and backward free energy for each window are taken as indicators of the consistency of the simulation.

An alternative method is that of thermodynamic integration in which the number of steps is radically increased such that the free energy, ΔA, can be approximated by

$$\Delta A = \int_0^1 \langle \Delta E(q^N) \rangle_\lambda \, d\lambda$$

where ΔE is the difference between the successive λ states but the average is taken with respect to the λ probability distribution. Thermodynamic integration can be carried out during the simulation itself (Berendsen *et al.*, 1985) using a method called 'slow growth' in which the coupling parameter λ is changed a very small amount at each dynamics time step. Comments on how λ should be varied and applications are given in Straatsma (1987), Straatsma *et al.* (1986), Straatsma and Berendsen (1988) and Pearlman and Kollman (1989). This technique is also implemented in the major distributed molecular dynamics programs.

Other methods of estimating free energy differences from simulation are the potential of mean force method (Pangali *et al.*, 1979; Chandrasekar *et al.*, 1984) and the probability ratio method (Bennett, 1976; Mezei *et al.*, 1985). These methods have not yet been used to study systems of biomolecules. However, very recent work suggests that the probability ratio method may be more efficient than the perturbation method under certain circumstances (Saito and Nakamura, 1990) and it may become more popular. Further details and discussion of each of these techniques may be found in Mezei and Beveridge (1986) and Beveridge and Dicapua (1989a, b).

As in all biomolecular simulations a suitable force field must be used to describe the solute and solvent. Commonly used force fields for proteins and organic molecules are the OPLS parameters (Jorgensen and Tirado-Rives, 1988) and those from AMBER (Weiner *et al.*, 1984). Current models for aqueous solvent are SPC (Berendsen *et al.*, 1981), TIP3P and TIP4P (Jorgensen *et al.*, 1983).

APPLICATIONS

Amino Acid Substitutions

An important goal of free energy simulations is to be able to predict the effect of site-specific mutations on the structure and stability of proteins. If a mutation is on the exterior surface of the protein, a major component of the free energy difference will involve interaction with the surrounding solvent. In order to provide the necessary basis for studying these effects in proteins, considerable effort has been expended in trying to understand the free energy of solvation of amino acid side chains.

Two extensive studies by Kollman and colleagues, using a molecular dynamics implementation of the free energy perturbation method (AMBER), have considered transformations of amino acids' side chains (Singh *et al.*, 1987; Bash *et al.*, 1987b). Both of these papers conclude that there is good agreement between experimental and simulation results and that the methodology may be extended to study single-residue mutations in proteins with some confidence. However, it is noted that some mutations fare much better than others and that there are unresolved problems with simulations of charged side chains, owing to the long-range nature of electrostatic interactions.

Recently we have begun to calculate the effect of the local environment on free energies of hydration (Saqi and Goodfellow, 1990; Goodfellow *et al.*, 1989). Our assumption is that, for a given chemical substitution (e.g. CH_3 to OH), the magnitude of the free energy change will depend on the local protein sequence and conformation. For example, one would expect the free energy change of the CH_3 to OH substitution in the mutation of ethane to methanol to be different to the identical substitution involved in the mutation of valine to threonine within a protein. Thus, we have calculated the free energy change for CH_3 to OH substitutions in three different environments.

We have used the program BOSS (W. L. Jorgensen, Purdue University, Indiana) which is a Monte Carlo protocol originally designed to simulate organic molecules in solution. This program uses the OPLS force field for the amino acids and the TIP4P model for the solvent. The calculations are carried out using the free energy perturbation method using a number of thermodynamic windows which go from $\lambda = 0$ (—CH3) to $\lambda = 1$ (—OH).

The three systems we have studied so far include the standard ethane to methanol mutation and then valine to threonine within two different helical peptides: ala–val–ala and ala–lys–val–lys–ala. A summary of our results is given in Table 6.1 from which it can be seen that the total free energy decreases from -8.3 ± 0.32 kcal mol^{-1} for the ethane to methanol mutation to -7.05 ± 0.18 kcal mol^{-1} for the valine to threonine mutation

Table 6.1 Summary of results for CH_3 to OH substitutions (in kcal mol^{-1})

System	*Forward*	*Backward*	*Average*
Ethane to methanol	8.68	−7.98	8.3 ± 0.32
val to thr I[a]	6.85	−7.20	7.05 ± 0.18
val to thr II[b]	6.79	−6.43	6.61 ± 0.18

[a] Valine to threonine mutation within the sequence ala–val–ala.
[b] Valine to threonine mutation within the sequence ala–lys–val–lys–ala.

in ala–val–ala and finally to -6.61 ± 0.18 kcal mol^{-1} for valine to threonine mutation within ala–lys–val–lys–ala.

We also observe that, despite hysteresis problems with individual windows, the overall free energy changes for the forward and backward runs are in reasonable agreement. A similar effect has been observed by others (Pettitt, 1989; Pearlman and Kollman, 1989) using slow growth methods. However, some windows, especially in the more complicated systems showed significant hysteresis and an obvious route is to refine the free energy procedure by subdividing this part of the calculation into more windows. Pearlman and Kollman (1989) have introduced a method of dynamically modifying windows.

We are also interested in the accuracy of these free energy perturbation methods. The use of thermodynamic windows allows us to calculate batch averages and standard deviations for each window from which we can calculate the total free energy change and its standard deviation. Such simple calculations lead to large but possibly more realistic error estimates than just the comparison of the total forward to backward free energy change. We have also looked at the effect of other variables in the protocol such as using two different starting seeds and/or changing to a constant pressure rather than a constant volume calculation.

Drug Design (Free Energy of Binding)

The design of molecules with specific biochemical action, based on knowledge of the structure of biological macromolecules, has become an important strategy for the design of new drugs (Hol, 1986). The current methods of 'rational drug design' are basically empirical; an amalgam of structural information and knowledge of the behaviour of simpler organic systems is used to design possible inhibitors. These inhibitors are then synthesized and their activities assessed. The process of synthesis is often difficult, time consuming and expensive. It would be a great advantage if the efficacy of the design could be tested without actually having to make the molecule. Computer simulations of the protein–drug interaction offer this possibility (McCammon, 1987).

One of the requirements for a good inhibitor is that it should have a high binding affinity (K), i.e. a large bias of the binding equilibrium in favour of the bound over the unbound state. The binding affinity is given by

$$K = \exp(-\Delta B/kT)$$

where ΔB is the difference between the free energy of the drug and protein when completely separate and when complexed. This free energy change could be estimated by slowly moving the two molecules together in small steps and using either thermodynamic integration or the perturbation method to find the sum of the free energy changes over the path. To obtain accurate results, for such a large structural change, a very large number of steps would be needed and this might be prohibitively expensive in computer time. We would also need detailed knowledge of any major structural changes of the protein which occur on drug binding.

In the practical rational design situation, often there is a model for the receptor and a series of possible inhibitors based on chemical and structural reasoning. We do not need to know the absolute free energy of the binding of each drug. Since it is only necessary to know which of the possible drugs has the highest affinity, we only have to calculate the difference in free energy of binding between pairs of similar drugs ($\Delta\Delta B$). The latter quantity can be calculated along a computationally simpler non-physical pathway using a Tembe–McCammon cycle (Tembe and McCammon, 1984). We can also assume that, for two similar ligands, any difference in protein conformation is localized to the active site, thus avoiding consideration of long-range effects.

The calculation has two stages. First, one drug is mutated into the other alone in solution and the relative free energy of hydration is calculated. Secondly, the same mutation is made but with the drug bound in the enzyme complex. Formally, the method is illustrated by the following thermodynamic cycle:

$$\begin{array}{ccc} \mathrm{E+S} & \xrightarrow{\Delta G_1} & \mathrm{ES} \\ \downarrow \Delta G_3 & & \downarrow \Delta G_4 \\ \mathrm{E+S'} & \xrightarrow{\Delta G_2} & \mathrm{ES'} \end{array}$$

The desired free energy change is given by $\Delta\Delta B = \Delta G_2 - \Delta G_1$. However, using the thermodynamic cycle, this is equivalent to $\Delta\Delta B = \Delta G_4 - \Delta G_3$, because free energies are state functions.

The first application of this technique to biological molecules was by Wong and McCammon (1986). They computed the relative free energies of binding of two benzamidine inhibitors to trypsin and of benzamidine to native and mutant trypsin. The mutant trypsin is one in which gly 216 was replaced by ala. The simulations were carried out at constant volume using the GROMOS package. The simulations were in agreement with experiment in indicating that benzamidine binds more strongly to trypsin than *p*-fluorobenzamidine, with $\Delta\Delta B = 3.8$ kJ mol^{-1} (0.91 kcal mol^{-1}), compared with an experimental value of 2.1 kJ mol^{-1} (0.50 kcal mol^{-1}). The errors were estimated to be about 1.1 kJ mol^{-1} (0.26 kcal mol^{-1}) in the simulation and about 0.65 kJ mol^{-1} (0.16 kcal mol^{-1}) in the experimental.

Experimental and theoretical results were also in agreement in that they both showed that benzamidine binds more strongly to trypsin than to the mutant of trypsin formed by the replacement of gly 216 by ala. The theoretical $\Delta\Delta B$ in this case was 5.6 kJ mol^{-1} (1.34 kcal mol^{-1}) while the experimental $\Delta\Delta B$ was 8.4 kJ mol^{-1} (2.01 kcal mol^{-1}). The uncertainties for both the theoretical and the experimental result were estimated to be larger than in the previous case.

Bash *et al.* (1987a) have studied two inhibitors bound to thermolysin. The perturbation is performed using the X-ray coordinates for one complex and by using model coordinates for one complex. Interestingly, the calculation in which the model coordinates are used leads to much smaller errors (i.e. 0.06 kcal mol^{-1} compared with 0.54 kcal mol^{-1}), where the binding free energy difference is around 4.2 kcal mol^{-1}. These errors are smaller than those found in Wong and McCammon's trypsin study. Both studies involve similar numbers of time steps. The lower error when model coordinates are used may imply that the configurations have been trapped in some local region of configuration space.

A study by Pettitt (1989) of the binding of drugs to the common cold virus (which is a member of the picornavirus family) gives values of -0.5 ± 3.0 kcal mol^{-1}. Thus, within the error, agreement is obtained with the experimental result of $+1.5$ kcal mol^{-1}. The perturbation method has also been used by Brooks (1989) to study the binding of four different inhibitors of dihydrofolate reductase. Experimental data are only quoted for one pair of inhibitors and this is not in agreement with the simulation, i.e. an experimental value of -1.7 kcal mol^{-1} compared with an estimate of $+1.9 \pm 0.7$ kcal mol^{-1}. Hirono and Kollman (1990) have studied the binding of 2′GMP and 2′AMP to ribonuclease T_1.

Similar methods can be used to compute other properties that are relevant to drug design. Recently, Essex *et al.* (1989) have calculated the relative free energy for two molecules transferred from water to carbon tetrachloride using the following thermodynamic cycle:

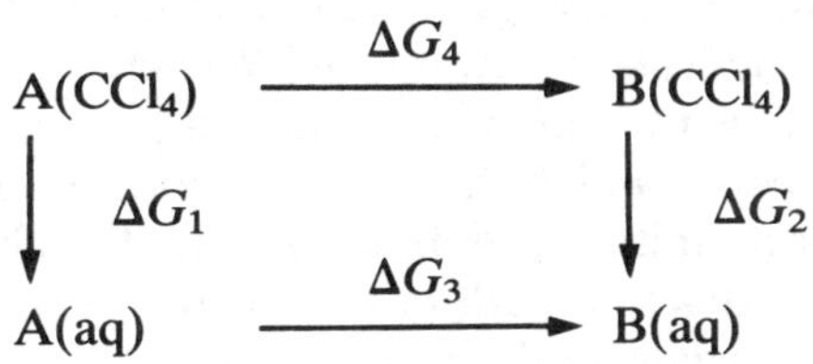

Thus, ΔG_4 and ΔG_3 are calculated in order to obtain an estimate of the difference $(\Delta G_1 - \Delta G_2)$ which gives the relative equilibrium distribution of A and B between the two solvents. The CCl_4 is intended as a simplified model for the properties of a membrane. This information about membrane solubility and thus the transport properties of a drug molecule would be of considerable use in drug design (Martin, 1978).

Enzyme Catalysis

Another important area of experimental interest is the modification of proteins by site-directed mutagenesis in order to modify the catalytic behaviour of enzymes. The rate of a catalytic reaction may be limited by either the rate of diffusional encounter between the reagents or the catalytic activation energy.

Initially, the reagents may move together as a result of random collisions with the surrounding water molecules. This rate of encounter is often enhanced by the electrostatic field of the enzyme, which attracts and orients the other reagents. This electrostatic field can be altered by site-directed mutagenesis and the changes in reaction rate calculated using the technique of Brownian reaction dynamics (Sharp *et al.*, 1987).

The rate may then be limited by the activation energy required for the catalysed reaction which can be modified by altering side chains within the active site of the enzyme. The free energy perturbation method can once again be used to calculate the effect of such changes by estimating the difference in the activation free energies of the native and mutant enzymes.

Two studies have been undertaken of mutations in subtilisin, since large changes in activity are known to occur for certain mutations. Rao *et al.* (1987) used the thermodynamic cycle approach to mutate asn 155 to ala. In this calculation a model for the transition state was obtained from quantum mechanical calculations of tetrahedral adducts, i.e. from gas phase information. Thus, this work assumes that this tetrahedral intermediate is similar to the transition state and that the transition state is similar for the enzyme before and after the mutation. These results show good agreement with experiment.

This transfer of gas phase structures to an aqueous system has been criticized by Warshel *et al.* (1988) who have therefore used a quite different approach. In this study, empirical valence bond theory has been used to show that the making and breaking of bonds in the reaction may be approximated by smoothly changing the potentials of the initial unreacted state to the final product state (Hwang and Warshel, 1987). The transition state of the molecule is identified as being that part of the path (between initial and final states) in which the configurations of the system have the same energy in both initial and final potentials. This perturbation is performed along the reaction path of both native and mutant enzymes in order to obtain the free energy difference. Initially it might appear that this reaction path perturbation and subtraction technique would be more expensive and less accurate than the thermodynamic cycle technique, but they compare both techniques (using their potentials), and find them to be very similar (Warshel *et al.*, 1988). Thus, both of these studies give similar results for subtilisin and good agreement with experimental data.

There is also considerable interest from both experimentalists and theoreticians in modifying the pH activity of enzymes by alteration of the environment of the ionizable active site residues (Russell and Fersht, 1987; Gilson and Honig, 1987; Sternberg *et al.*, 1987). The current approach has tended to involve the introduction of other charged residues by site-directed mutagenesis into the enzyme followed by calculation of changes to the electrostatic field using macroscopic theories. The free energy perturbation method should allow more subtle changes to be modelled at the atomic level. However, so far only the pK_a values of small organic molecules have been calculated (Jorgensen *et al.*, 1987). Free energy calculations have also been used to calculate the redox potentials of potential bioreductive agents (Reynolds *et al.*, 1988).

Conformational Changes

Proteins

There has been considerable study of the possible conformations of the alanine dipeptide, a typical protein backbone unit, which together with several of its analogues have been shown to act as potent anti-epileptics (Conley and Kohn, 1987). Initial calculations were carried out *in vacuo* using both quantum and molecular mechanics and showed that the C_7 conformation is the most stable (Weiner and Kollman, 1981). Experimental results (Madison and Kopple, 1980) and theory agree that the C_7 conformation is the most stable in non-polar solvents but conflicting and inconclusive results exist on the structure in water. Mehrotra *et al.* (1984), on the basis of internal energy simulations, suggested that the α_R and P_{11} configurations are preferentially stabilized by water. Mezei *et al.* (1985) estimated differences in free energies of hydration for the four conformations, using Monte Carlo umbrella sampling methods. These results indicate a preference for both the α_R and P_{11} forms by about 3.5 kcal mol^{-1}, over the C_7. The enthalpic and entropic terms are found to be of similar importance. Anderson and Hermans (1988) have recently completed a more extensive study of the conformations of alanine dipeptide using a molecular dynamics perturbation technique, producing a 'free energy Ramachandran plot'. These results clearly show the order of stability is the β, α, C_7 conformations with C_7 being the most stable.

The free energy difference between *trans* and *cis* forms of the peptide bond has been studied by Jorgensen and Gao (1988). They consider the difference in the gas phase using *ab initio* quantum mechanics and then in aqueous solution using Monte Carlo perturbation techniques and conclude that the observed *trans* preference is entirely due to the gas phase intramolecular entropy.

There have been several attempts at conformational studies of larger peptide fragments (Hagler *et al.*, 1979; Scheraga, 1971; Scheraga and Li, 1988; Paine and Scheraga, 1987; Ripoli and Scheraga, 1988; Meirovitch *et al.*, 1987). These studies are carried out *in vacuo*. Although chains of up to 19 residues can be handled using a potential energy approach (Ripoli and Scheraga, 1988), even *in vacuo* free energy techniques can handle only full conformational searches of pentapeptides.

Nucleotides

In contrast to the large number of studies on protein, there are relatively few on free energy differences associated with changes in composition of a ligand binding to nucleotides. This may reflect the more 'complex' nature of nucleotides from the computational point of view. The nucleotide backbone is relatively flexible. Moreover, the net charge on each phosphate group leads to large electrostatic energies and this net charge is balanced by counterions in the surrounding solvent. Thus, modelling nucleotides in solution at the atomic level becomes a complex problem. Jayaram *et al.* (1988a,b) have studied the solvent-related free energy changes associated with different conformers of the dimethyl phosphate anion and found that the *gauche–gauche* conformation is the most stable. Free energy estimates for the mutations of methane to adenine and methane to thymine are available as well as thymine to cytosine and adenine to guanine (Bash *et al.*, 1987b). These calculations have been extended to study the association of bases *in vacuo* and in water. It was found that stacked complexes are slightly more stable than hydrogen-bonded complexes in water but *in vacuo* the hydrogen-bonded complexes are favoured (Cieplak *et al.*, 1988).

Further studies by Kollman's group are in progress on the effect of methylation of guanine on the B to Z transition, the relative binding of netropsin and 'ethatroprosin' and sequence dependence of daunomycin binding to DNA.

SUMMARY

The examples described above indicate the importance of free energy perturbation calculations for rational engineering of proteins and design of drugs as well as a better understanding of the conformational stability of macromolecules. However, it is clear that there are still problems with the accuracy and precision of these calculations. In the future, it will be necessary to improve the accuracy of estimates for free energy changes and to extend the range of this technique so that we can study larger changes.

ACKNOWLEDGEMENTS

We wish to thank the Science and Engineering Research Council for a studentship to M.A.W. and for support under project grants GR/E72737, ASL7802, GR/86683 and GR/E47025. J.M.G. thanks the Wellcome Trust for a research leave fellowship.

APPENDIX 1: ENERGY, ENTROPY AND FREE ENERGY

In order to understand fully the theoretical background to the applications of free energy calculations it is necessary to have a knowledge of statistical mechanics. This area of physics is concerned with the behaviour of a large number of particles. One of the basic assumptions is that if any system (e.g. molecule) can obtain a maximum energy E from its surroundings it is equally likely to be in any of its energy states up to this maximum.

If an ensemble of such molecules is given an approximate total energy E_{TOT} by its surroundings, the systems arrange the energy randomly among themselves. The instantaneous arrangement of energy between the molecules is not important as all we can observe is the overall behaviour of the ensemble of molecules. For example, the distribution of energies (i.e. the number of molecules having a particular energy) may determine the rate of a reaction. A particular distribution can be produced by very many different microscopic arrangements of the energy and the most probable distribution of the energy is the one that can be produced in the largest number of ways. This most probable distribution (called the Boltzmann distribution) is such that the probability of a molecule's having a particular energy is proportional to exp(−energy/kT) where T is the temperature and k the Boltzmann constant. Fortunately, it turns out that this distribution is so much more likely for large ensembles than any other distribution that it is the actual observed energy distribution.

In order to consider the relationship between free energy and the distribution of energies described above it is useful to consider a simple example in which a molecule may exist in three forms (A, B and C) in solution. At equilibrium, there will be a certain proportion of each in solution. Let us consider many solute molecules in a large amount of water to constitute one ensemble and each individual solute plus its closest waters to be a system. In this case, each system has many possible internal configurations and thus many possible energy states which are represented by their density of states (number of configurations per unit energy having a given energy). In this example, the ensemble of one of the forms (A, B, or C) will have the greatest number of arrangements of energy and thus the highest entropy.

In order to see which states of each form (A, B or C) would be occupied at equilibrium, the number of states available at each energy for each form is multiplied by the probability of a system's having that energy (i.e. exp(−energy/kT)). If most of the states of B are of lower energy than A then B will have many more occupied states than A. Thus, B has a high probability and most of A converts to B. If we then look at the equilibrium between B and C we may find that C can have many more occupied states because C has greater entropy (many more energy states). It is also necessary to remember that because the ratio of A:B:C depends on the Boltzmann factor it will change with temperature.

In summary, the systems of an ensemble will be distributed among their possible energy states in such a way as to maximize the number of possible arrangements. If the system can exist in two or more forms, the number of systems adopting each form is determined by the energy and entropy of each of the forms. Mathematically, we can write the change in free energy ΔA as

$$\Delta A = -kT \ln\langle \int \rho(E) \exp(-E/kT)\, \mathrm{d}E \rangle$$

where $\rho(E)$ represents the density of states at energy E. There is an alternative representation in which the integration can be over the number of possible configurations rather than over the possible energy states. Further mathematics can relate this equation from statistical mechanics to the classical thermodynamic equation in which

$$\Delta G = \Delta H - T\,\Delta S$$

where ΔH represents the change in enthalpy (which can be equated to the internal energy at constant volume) and ΔS represents the change in entropy between two states. ΔG rather than ΔA implies conditions of constant numbers of particles, pressure and temperature.

Further discussion of this molecular view of entropy and free energy can be found in Kauzmann (1967) and more derivations are given in Huang (1969).

REFERENCES

Anderson, A. G. and Hermans, J. (1988) Microfolding: conformational probability for alanine dipeptide in water from molecular dynamics simulations. *Proteins*, **3**, 262–265

Bash, P. A., Singh, U. C., Brown, F. K., Langridge, R. and Kollman, P. A. (1987a). Calculation of relative change in binding free energy of a protein-inhibitor complex. *Science*, **235**, 574–575

Bash, P. A., Singh, U. C., Langridge, R. and Kollman, P. A. (1987b). Free energy calculations by computer simulation. *Science*, **236**, 564–565

Bennett, C. H. (1976). Efficient estimation of free energy differences from Monte Carlo data. *J. Comput. Phys.*, **22**, 245–268

Berendsen, H. J. C., Postma, J. P. M., van Gunsteren, W. F. and Hermans, J. (1981). Interaction models for water in relation to protein hydration. In Pullman, B. (Ed.), *Intermolecular Forces*. Reidel, Dordrecht, pp. 331–342

Berendsen, H. J. C., Postma, J. P. M. and van Gunsteren, W. F. (1985). In Hermans, J.

(Ed.), *Molecular Dynamics and Protein Structure*. Polycrystal Book Service, Illinois, pp. 43–46

Beveridge, D. L. and Dicapua, F. M. (1989a). Free energy via molecular simulation: a primer. In van Gunsteren, W. F. and Weiner, P. (Eds.), *Computer Simulation of Biomolecular Systems*. ESCOM, Leiden, pp. 1–26

Beveridge, D. L. and Dicapua, F. M. (1989b). Free energy via molecular simulation: applications to chemical and biomolecular systems. *Annu. Rev. Biophys. Biophys. Chem.*, **18**, 431–492

Brooks, B. R., Bruccoleri, R. E., Olafson, B. D., States, D. J., Swaminathan, S. and Karplus, M. (1983). CHARMM: a program for macromolecular energy minimization and dynamics calculations. *J. Comput. Chem.*, **4**, 187–217

Brooks, C. L. (1989). Thermodynamic calculations on biological molecules. In van Gunsteren, W. F. and Weiner, P. (Eds.), *Computer Simulation of Biomolecular Systems*. ESCOM, Leiden, pp. 73–88

Chandrasekar, J., Smith, S. F. and Jorgensen, W. L. (1984). S_n2 reaction profiles in gas phase and aqueous solution. *J. Am. Chem. Soc.*, **106**, 3049–3050.

Cieplak, P. and Kollman, P. A. (1988). Calculation of the free energy of association of nucleic acid bases in vacuo and water solution. *J. Am. Chem. Soc.*, **110**, 3734–3739

Conley, J. D. and Kohn, H. (1987). Functionalised DL-amino acid derivatives: potent new agents for the treatment of epilepsy. *J. Med. Chem.*, **30**, 567–574

Creighton, S., Hwang, J., Warshel, A., Parson, W. W. and Norris, J. (1988). Simulating the dynamics of the primary charge separation process in bacterial photosynthesis. *Biochemistry*, **27**, 774–781

Essex, J. W., Reynolds, C. A. and Richards, W. G. (1989). Relative partition coefficients from partition functions: a theoretical approach to drug transport. *J. Chem. Soc., Chem. Commun.*, 1152–1154

Gilson, M. K. and Honig, B. H. (1987). Calculation of electrostatic potentials in an enzyme active site. *Nature*, **330**, 84–86

Goodfellow, J. M., Saqi, M. A. S., Thanki, N., Baum, J. O. and Finney, J. L. (1989). Monte Carlo calculations on proteins: hydration and electron transfer processes. In Beveridge, D. L. and Lavery, R. (Eds.), *Theoretical Chemistry and Molecular Biophysics*. Adenine Press, New York

van Gunsteren, W. F. (1989) Methods for calculations of free energies and binding constants: successes and problems. In van Gunsteren, W. F. and Weiner, P. (Eds.), *Computer Simulation of Biomolecular Systems*. ESCOM, Leiden, pp. 27–59

Hagler, A. T., Stern, P. S., Sharon, R., Becker, J. M. and Naider, F. (1979). Computer simulation of the conformational properties of oligopeptides. *J. Am. Chem. Soc.*, **101**, 6842–6852

Hermans, J., Berendsen, H. J. C., van Gunsteren, W. F. and Postma, J. P. M. (1984). A consistent empirical potential for water protein interactions. *Biopolymers*, **23**, 1513–1518

Hirono, S. and Kollman, P. A. (1990). Calculation of the relative binding free energy of 2′GMP and 2′AMP to ribonuclease T, using molecular dynamics/free energy perturbation approaches. *J. Mol. Biol.*, **212**, 197–209

Hol, W. G. J. (1986). Protein crystallography and computer graphics—towards rational drug design. *Angew. Chem., Int. Edn. Engl.*, **25**, 767–778

Huang, K. (1969). *Statistical Mechanics*. Wiley, New York

Hwang, J.-K. and Warshel, A. (1987). Semiquantitative calculations of catalytic free energies in genetically modified enzymes. *Biochemistry*, **26**, 2669–2673

Jayaram, B., Mezei, M. and Beveridge, D. L. (1988a). Conformational stability of dimethyl phosphate anion in water: liquid state free energy simulations. *J. Am. Chem. Soc.*, **110**, 1691–1694

Jayaram, B., Ravishkanker, G. and Beveridge, D. L. (1988b). Conformational preferences of phosphodiester torsional angles in dimethylphosphate anion in free space. *J. Phys. Chem.*, **92**, 1032–1034

Jorgensen, W. L. and Gao, J. (1988). *Cis–trans* energy difference for the peptide bond in the gas phase and in aqueous solution. *J. Am. Chem. Soc.*, **110**, 4212–4216

Jorgensen, W. L. and Ravimohan, C. (1985). Monte Carlo simulation of differences in free energies of hydration. *J. Chem. Phys.*, **83**, 3050–3054

Jorgensen, W. L. and Tirado-Rives, J. (1988). The OPLS potential functions for proteins. *J. Am. Chem. Soc.*, **110**, 1657–1666

Jorgensen, W. L., Chandrasekhar, J., Madura, J. D., Impey, R. W. and Klein, M. L. (1983). Comparison of simple potential functions for simulating liquid water. *J. Chem. Phys.*, **79**, 926–935

Jorgensen, W. L., Briggs, J. M. and Gao, J. (1987). *A priori* calculations of pK_as for organic compounds in water. *J. Am. Chem. Soc.*, **109**, 6857–6858

Kauzmann, W. (1967). *Thermodynamics and Statistics: with Applications to Gases*. Benjamin, Menlo park, CA

Lybrand, T. P., Ghosh, I. and McCammon, J. A. (1985). Hydration of chloride and bromide anions: determination of relative free energy by computer simulation. *J. Am. Chem. Soc.*, **107**, 7793–7794

Madison, V. and Kopple, K. D. (1980). Solvent-dependent conformational distributions of some dipeptides. *J. Am. Chem. Soc.*, **102**, 4855–4863

Martin, Y. C. (1978). *Quantitative Drug Design*: 1, Dekker, New York

McCammon, J. A. (1987). Computer-aided molecular design. *Science*, **238**, 486–491

McCammon, J. A. and Harvey, S. (1987). *Dynamics of Proteins and Nucleic Acids*. Cambridge University Press, Cambridge

Mehrotra, P. K., Mezei, M. and Beveridge, D. L. (1984). Monte Carlo determination of the internal energies for the alanine dipeptide in the C_7, C_5, α_R and P_{II} conformations. *Int. J. Quantum Chem., Quantum Biol. Symp.*, **II**, 301–308

Meirovitch, H., Vasquez, M. and Scheraga, H. A. (1987). Stability of polypeptide conformational states as determined by computer simulation of the free energy. *Biopolymers*, **26**, 651

Metropolis, N., Rosenbluth, A. W., Rosenbluth, M. N., Teller, A. H. and Teller, E. (1953). Equation of state calculations by fast computing machines. *J. Chem. Phys.*, **21**, 1087–1092

Mezei, M. (1982). Comparisons of free energy of different water models computed by Monte Carlo methods. *Mol. Phys.*, **47**, 1307–1315

Mezei, M. and Beveridge, D. L. (1986). Free energy simulations. *Ann. N.Y. Acad. Sci.*, **482**, 1–12

Mezei, M., Mehrotra, P. K. and Beveridge, D. L. (1985). Monte Carlo determination of the free energy and internal energy of hydration for the alanine dipeptide at 25°C. *J. Am. Chem. Soc.*, **107**, 2239–2245

Paine, G. H. and Scheraga, H. A. (1987). Prediction of the native conformation of a polypeptide by a statistical–mechanical procedure. *Biopolymers*, **26**, 1125–1162

Pangali, C., Rao, M. and Berne, B. J. (1979). Monte Carlo simulation of the hydrophobic interaction. *J. Chem. Phys.*, **71**, 2975–2981

Pearlman, D. and Kollman, P. A. (1989). A new method for carrying out free energy perturbation calculations: dynamically modified windows. *J. Chem. Phys.*, **90**, 2460–2470

Pettitt, B. M. (1989). Successes, failures and curiosities in free energy calculations. In van Gunsteren, W. F. and Weiner, P. (Eds.), *Computer Simulation of Biomolecular Systems*. ESCOM, Leiden, pp. 94–100

Quirke, N. (1980). In *Proceedings of NATO Summer School on Superionic Conductors, Odense*. Plenum, New York

Rao, S. N., Singh, U. C., Bash, P. A. and Kollman, P. A. (1987). Free energy perturbation calculations on binding and catalysis on mutating Asn 155 in subtilisin. *Nature*, **328**, 551–554

Reif, F. (1965). *Fundamentals of Statistical and Thermal Physics*. McGraw-Hill, New York

Reynolds, C. A., King, P. M. and Richards, W. G. (1988). Computed redox potentials and the design of bioreductive agents. *Nature*, **334**, 80–82

Ripoli, D. R. and Scheraga, H. A. (1988). On the multiple-minima problem in conformational analysis of polypeptides. *Biopolymers*, **27**, 1283–1303

Russell, A. J. and Fersht, A. R. (1987). Rational modification of enzyme catalysis by engineering surface charge. *Nature*, **328**, 496–500

Saito, M. and Nakamura, H. (1990). Hydration free energy calculations by the acceptance ratio method. *J. Comput. Chem.*, in press

Saqi, M. A. S. and Goodfellow, J. M. (1990). Free energy changes associated with amino acid substitution in proteins. *Protein Eng.*, in press

Scheraga, H. A. (1971). Theoretical and experimental studies of conformations of polypeptides. *Chem. Rev.*, **71**, 195–217

Scheraga, H. A. and Li, Z. (1988). Structure and free energy of complex thermodynamic systems. *J. Mol. Struct. (Theochem.)*, **179**, 333–352

Sharp, K., Fine, R. and Honig, B. (1987). Computer simulations of diffusion of a substrate to an active site of an enzyme. *Science*, **236**, 1460–1463

Singh, U. C., Brown, F. K., Bash, P. A. and Kollman, P. A. (1987). An approach to the application of free energy perturbation methods using molecular dynamics. *J. Am. Chem. Soc.*, **109**, 1607–1614

Sternberg, M. J. E., Hayes, F. R. F., Russell, A. J., Thomas, P. G. and Fersht, A. R. (1987) Prediction of electrostatic effects of engineering of protein charges. *Nature*, **330**, 86–88

Straatsma, T. P. (1987). Free energy evaluation by molecular dynamics simulations. *Ph.D. Thesis*, University of Groningen

Straatsma, T. P. and Berendsen, H. J. C. (1988). Free energy of ionic hydration: analysis of a thermodynamic integration technique to evaluate free energy differences by molecular dynamics simulations. *J. Chem. Phys.*, **89**, 5876–5886

Straatsma, T. P., Berendsen, H. J. C. and Postma, J. P. M. (1986). Free energy of hydrophobic hydration: a molecular dynamics study of noble gases in water. *J. Chem. Phys.*, **85**, 6720–6727

Sussman, F., Goodfellow, J. M., Barnes, P. and Finney, J. L. (1985). Calculation of free energy differences for water from computer simulation. *Chem. Phys. Lett.*, **113**, 372

Tembe, B. L. and McCammon, A. J. (1984). Ligand–receptor interactions. *Comput. Chem.*, **8**, 281–283

Torrie, G. M. and Valleau, J. P. (1977). Non-physical sampling distributions in Monte Carlo free energy estimation: umbrella sampling. *J. Comput. Phys.*, **23**, 187–199

Warshel, A., Sussman, F. and Hwang, J. K. (1988). Evaluation of catalytic free energies in genetically modified proteins. *J. Mol. Biol.*, **201**, 139–159

Weiner, P. K. and Kollman, P. A. (1981). Quantum and molecular mechanical studies on alanyl dipeptide. *J. Comput. Chem.*, **2**, 287

Weiner, S. J., Kollman, P. A., Case, D. A., Singh, U. C., Ghio, C., Alagona, G., Profeta, S. and Weiner, P. (1984). A new force field for molecular mechanical simulation of nucleic acids and proteins. *J. Am. Chem. Soc.*, **106**, 765–784

Wong, C. F. and McCammon, J. A. (1986). Dynamics and design of enzymes and inhibitors. *J. Am. Chem. Soc.*, **108**, 3830–3832

Index